Contemporary Marketing Management for Tourism and Hospitality

Nikolaos Stylos • Roya Rahimi
Peter Robinson
Editors

Contemporary Marketing Management for Tourism and Hospitality

How New Tech and Gen Z Drive Sustainable Business Transformation

palgrave
macmillan

Editors
Nikolaos Stylos
University of Bristol Business School
Bristol, UK

Roya Rahimi
University of Wolverhampton
Business School
Wolverhampton, UK

Peter Robinson
Leeds Beckett University
Leeds, UK

ISBN 978-3-031-65048-2   ISBN 978-3-031-65049-9 (eBook)
https://doi.org/10.1007/978-3-031-65049-9

This Palgrave Macmillan imprint is published by the registered company Springer Nature Switzerland AG.
The registered company address is: Gewerbestrasse 11, 6330 Cham, Switzerland

Preface

The marketing discipline is developing fast—it serves as the basis for new marketing streams such as sustainability, social and institutional marketing. The advent of new digital technologies and innovative applications and their penetration into marketing management has created new opportunities and challenges for marketing scholars and practitioners. It is therefore important to understand the different perspectives and norms that underpin current marketing theory and contemporary practice. This has necessitated a change in how marketing is perceived by organizations. New marketing streams, which are influenced by the latest advances in innovation and sustainability, and a completely new generation of consumers (i.e. GenZers) who were born in the Industry 4.0 era, shape new background and social trends.

This book offers opportunities for readers to engage with the latest theory and practice of marketing in tourism and hospitality. Tourism is a key driver for economic and social impact in both developed and developing countries. It has the power to underpin the viability of other industries and rejuvenate the business sector, as well as supporting national economies to mitigate the repercussions emerging from challenging geopolitical factors.

Previously published books have mainly focused on Web 2.0 technologies, such as social media platforms with reference to the generation of Millennials. This book concentrates on the most recent and contemporary topics of marketing in the tourism and hospitality industry. It introduces an overarching structure of three key dimensions: Digitalization; Innovation; and Sustainability across different generations, to address the most contemporary topics in these areas holistically. This is important as these three areas of work will be key pillars for marketing practice over the coming decades.

The book contextualizes these key issues within contemporary marketing practice, consumer behaviour, and the role of stakeholders. Scholars, managers, destination management organizations and others interested in or working in the tourism sector will find this book an essential reference point to deliver effective marketing practice and as a reference source for updating their knowledge. The book also fills the gap for a much-needed marketing text that recognizes the influential role of the three pillars mentioned above to produce state-of-the-art knowledge in Tourism and Hospitality Marketing Management. It is also one of the first books offering a collection of selected topics aiming to better understanding how new technologies, creativity, and sustainability shape the global tourism sector of our times. This should be of interest to managers, scholars, and entrepreneurs interested in acquiring a more concrete knowledge on the crucial role of the current marketing paradigm shift with respect to managing tourism and hospitality-related services. In academia the book covers the following areas of Marketing: Services Marketing; Tourism Marketing and Management; Hospitality Management; Event Management; and Business Management.

The editorial team would like to thank very much all authors contributing to this book and Palgrave Macmillan publishers for their invaluable support to materializing this publication.

We hope that you all find this a useful companion and reference point in the new world of tourism and hospitality marketing management.

Bristol, UK — Nikolaos Stylos
Wolverhampton, UK — Roya Rahimi
Leeds, UK — Peter Robinson

Contents

Notes on Contributors

Elizabeth Agyeiwaah is a lecturer at the UQ Business School. She is also a senior research associate at the School of Tourism and Hospitality, at the University of Johannesburg. She holds a PhD from the Hong Kong Polytechnic University. Her research interests include sustainable development in tourism and hospitality, small and medium tourism enterprises, and tourist studies.

Nina Balan holds a PhD in Planning and Urban Studies from the University of New Orleans, focused on sustainable preservation strategies and climate change impacts on coastal urban tourism. Her research has been published in journals like SHIMA and cited by the Urban Institute and RAND. With an MS in Tourism Management and a BS in Languages and Literature, Nina brings a multidisciplinary approach. At The Beach at UNO, she develops public-private partnerships, delivering sustainable solutions for unpredictable environments. Her current work explores ableism and disability accessibility planning in New Orleans. Nina serves von the boards of the local non-profit Heart N Hands and the American Planning Association New Orleans Metro Section. Her expertise spans urban planning, tourism, cultural preservation and creating inclusive, resilient cities.

Shelley Beck is Senior Lecturer in Enterprise and Entrepreneurship and the subject coordinator for the MSc Business and Management programme at Oxford Brookes University. Shelley obtained her DCom (Business Management) from Nelson Mandela University in South Africa in 2018 on her work that focused on the influence of parents on next-generation family members entering the family business. Her area of research focus is in entrepreneurship, particularly, family entrepreneurship and SMEs. She has been in tertiary education for more than 10 years and taught various modules in the areas of entrepreneurship, marketing and research. In that time, her work has been presented at several international business conferences and journals.

Zoe Charis Belenioti is Lecturer in Digital Marketing, YSJ London Campus. Since 2020, she has acquired a trans-European teaching experience in the field of digital marketing, marketing communications in healthcare, tourism marketing, social marketing and consumer behaviour. Her cross-disciplinary work, focusing on digital marketing, digital transformation, cultural tourism, social media, brand co-creation, marketing and management of cultural institutions and consumer behaviour, has been repeatedly cited (hi = 8, i10 = 7). Her research management skills were heavily challenged and mastered by her novel and broad PhD topic, which required a multidisciplinary and mixed-method approach.

She has been a research fellow in Digital Communication funded by the European Commission at the University of Copenhagen (2015). Moreover, she served as a fellow of the Stavros Niarchos Foundation, iDeas Lab and CSIS's Transnational Threats Project on disinformation's impact on EU elections. In between her studies, she worked across Europe as communications specialist at the European Commission in Brussels (DGCNECT), the Principal Directorate of Communication at the European Patent Office in Munich, and for more than ten years as press office assistant at Thessaloniki International Film Festival. She also served as Principal Investigator at the Social Inclusion Observatory, DG Public Health of Region Central Macedonia.

Cemile Ece is a PhD candidate in the Department of Tourism and Hotel Management at Eskişehir Osmangazi University and project expert. She currently researches eco-tourism, sustainability, diversity, equity, inclusion, destination management, destination branding and consumer behaviour. Her last research was supported by the Scientific

Research Centre of ESOGÜ. She has published scientific articles and projects on tourism. She has practical experience in the tourism field for more than 10 years as a professional. She worked in the front office and guest relations departments at multi-national hotels, and she has been working as a project expert in NGOs and takes part in NGO activities and takes an active role, especially on environmental, sustainable, ecological and inclusion issues. She has experience in EU+ projects and National Funded Projects.

Stefania Escobar is a research assistant and doctoral candidate in Economics and Management at the Free University of Bozen-Bolzano. She is a recipient of the European Master of Tourism Management (EMTM). She was a visiting research scholar at Free University of Brussels (ULB), New York University (USA) and University of Aveiro (Portugal). Her research lies at the intersection of management, smart cities, tourism and sustainability. Her research has been presented at TTRA, APAC-CHRIE, SSPCR EURAC and third World Conference on Smart Destinations. Her work has appeared in the journal *Cities*.

Efnan Ezenel is a PhD candidate in the Department of Tourism and Hotel Management at Eskişehir Osmangazi University and a licensed tourist guide of Turkey by the Turkish Ministry of Culture and Tourism. She currently researches inclusive tourism, accessibility, social inclusion, diversity, equity, inclusion, destination management, destination branding and consumer behaviour. She has also engaged in rural tourism studies. She has practical experience in the tourism field for more than 12 years as a professional. She worked as a guest relations manager at multi-national hotels in the hospitality sector. She takes an active part in disability, inclusion and accessibility issues and has been working for NGOs on disability and disability rights. Additionally, she has worked on EU+ projects and National Funded Projects.

Sofia Gkarane holds a PhD in Tourism Marketing from the Department of Business Administration at the University of Macedonia. She also holds a Master's in Business Administration and a Bachelor's degree in International Economics from the University of Macedonia. She is currently a post-doctoral researcher at the Department of Business

Administration at the University of Macedonia. Her post-doctoral research proposal is entitled "Cultural Heritage and Seasonality: Enhancing Sustainable Marine Tourism in Greece".

Her research interests include tourism management, event management, tourism seasonality, sports marketing and international marketing. She has participated in a number of conferences and co-authored several research papers and book chapters. She brings over 12 years of extensive experience in international business, specializing in product exports to a diverse range of global markets, including Asia, the Balkans, Europe and the Middle East. She actively contributes as a researcher in EU-funded research projects and also holds a position as a lecturer at NYC.

Mengyun Hu is Lecturer in Information Systems Management at Newcastle University. She holds a PhD in Innovative Technologies and Marketing. His research projects focus on technology adoption, services marketing with new technologies, tourism marketing and consumers' well-being.

Ioannis Katsanakis is an assistant professor at the Department of Tourism Studies at the University of Piraeus in the subject of Information Systems Management in Businesses and Organizations with an emphasis on Business Process Management.

Imran Maqbool Khan is Senior Lecturer in Marketing at Birmingham City Business School University. Prior to this, he was Senior Lecturer in Digital Marketing at the University of Wolverhampton, and also worked at the University of Salford as a Lecturer in Digital Marketing, and at University of York as a Lecturer in Marketing. He completed his PhD (Digital Marketing), MSc (Marketing) and Postgraduate Certificate of Academic Practice (PGCAP) from the University of Salford, UK. Khan is specialized in digital marketing and currently teaches digital marketing-related modules at both undergraduate and postgraduate levels. Khan has initiated WLV Marketing Bootcamp sessions where he invites marketing experts from the industry to share their stories and teach various digital marketing tools. Khan's research explores the impact of digital technology on human behaviour. He has published his research in a high impact factor journal. He has also presented in many national and international conferences. He has also provided research training to national and international universities. He is a lifetime member of the Digital Marketing Institute (DMI), Senior Fellow of Higher Education Academy (SFHEA) and Member of Academy of Marketing.

Christos Klintsaris is a master graduate of the Master of Tourism Business Administration at the Hellenic Open University (HOU). His master's thesis was about the contribution of digital marketing to hotel business.

Chang (Cherise) Li is a PhD candidate at the Faculty of Hospitality and Tourism Management, Macau University of Science and Technology. Her current research interests are dark tourism, sustainable tourism development, the workforce in tourism and hospitality, backpacker and intercultural study.

Alison Maton is an experienced tourism academic, associate professor and Head of Partnerships at Buckinghamshire New University. Her expertise lies in sustainable tourism, responsible tourism, overtourism and the SDGs. She possesses over a decade of teaching experience and has extensive experience in curriculum design and development. She has successfully authored and led several tourism courses at undergraduate and postgraduate levels, working closely with leading professional bodies and industry as part of this process. Furthermore, she is a senior fellow of the Higher Education Academy, a Certified Management and Business Educator and an advocate for widening participation in Higher Education.

Eleonora Pantano is Associate Professor (Reader) of Retail and Marketing Technology. Her research activities mainly relate to the development of new customer solutions and retail management strategies, with emphasis on the role of new technology, artificial intelligence and machine learning algorithms. Her findings appear in several books and textbooks and in international peer-reviewed journals (i.e. *Journal of Service Research*, *Annals of Tourism Research* and *British Journal of Management*).

Albena Pergelova is Professor of Marketing and currently holds the position of Board of Governors Research Chair at MacEwan University School of Business, Edmonton, Canada. Her research is interdisciplinary (entrepreneurship and marketing) with a focus on social and emancipatory aspects in entrepreneurship, consumer well-being, and digital technologies. Her research has been published in leading international journals across different fields (e.g. *Journal of Business Ethics, Journal of Business Research, Journal of Small Business Management, International Small Business Journal, Entrepreneurship & Regional Development, Journal of Advertising, Journal of Consumer Affairs* and *Journal of Marketing Management*, among others) and recognized with numerous international awards (e.g. BCERC Best Paper in Social Entrepreneurship, DIANA Conference Best Paper in Women Entrepreneurship, several Best Paper Awards from the Academy of Marketing, etc.)

Roya Rahimi is a Reader in Marketing and Leisure Management. She is the REF coordinator of UoA 17 at the University of Wolverhampton, Business School. Her research interests are consumer experience, innovation, big data, sharing Economy, CRM, organizational culture, gender equality and tourism higher education. She has more than 50 academic publications including 3* and 4* journals with high impact factors. Her work has been published in top-tier journals such as *Annals of Tourism Research, Journal of Tourism and Hospitality Research, Journal of Travel & Tourism Marketing, International Journal of Contemporary Hospitality Management* and *Anatolia*. Her work has been also presented at various international conferences and appears in book chapters released by Routledge, CABI, Emerald, and IGI. She has been published in a variety of languages. Roya is the Associate Editor for *Journal of Tourism Management Perspectives*, book reviews editor for *Journal of Hospitality & Tourism Management* (JHTM) and book review editor for *Journal of Hospitality and Tourism Technology*. She sits on the editorial board of different leading journals including the *Journal of Hospitality and Tourism Technology, European Management Review, European Journal of Tourism Research, Journal of Hospitality & Tourism Management* and *The International Journal of Tourism Sciences*. Roya co-edited a special issue on sustainability in the Tourism and Hospitality Industry for the *International Journal of Contemporary Hospitality Management*.

Peter Robinson is the Head of the Centre for Tourism and Hospitality Management at Leeds Beckett University. He is the author of several books and peer-reviewed journals and has appeared on a number of television and radio programmes discussing the tourism industry. Peter has owned businesses in the travel, ICT, waste management and publishing sectors, and has delivered consultancy work for organizations in a wide range of sectors across the world. In 2021 Peter was Awarded the Victor Middleton Prize for Education and Scholarship by the Association for Tourism in Higher Education (ATHE). Peter is a Fellow of the Tourism Management Institute and the Yorkshire and Humber representative, a Fellow of the Tourism Society, a Fellow of the Institute of Travel and Tourism, and a Principal Fellow of the Higher Education Academy. He is Co-Chair of The Association for Tourism in Higher Education (ATHE) and a Director of The Institute of Travel and Tourism, and he represents ATHE on the Tourism Society Board. He is Chair of a £41 m restoration of a historic house and country park near Derby and was formerly Vice-Chair of The Museum of Carpet. His research interests focus on business planning, visitor experiences, heritage management and Cold War tourism. Previously, Peter worked as Head of Marketing, Leisure, Innovation and Enterprise at the University of Wolverhampton, and also managed the institution's Arts Council–funded Arena Theatre. Before working in academia Peter managed EU-funded projects in the Peak District, tourism projects in Oxfordshire, and spent a number of years working for The National Trust.

Daniele Scarpi has been Associate Professor of Marketing at the University of Bologna since 2014. His research activities relate to consumer behaviour. His papers have appeared in leading journals such as the *Journal of Service Research*, *Journal of Retailing*, *Journal of Interactive Marketing*, *Industrial Marketing Management*, *Tourism Management*, *Travel Research*, *Marketing Letters*, *Annals of Tourism Research* and several others.

Nikolaos Stylos is Associate Professor (Reader) of Marketing and Digital Innovation, *Innovation & Digitalisation* research group Lead and the *Smart Networks 4 Sustainable Futures* faculty research group Founder/Lead at the University of Bristol Business School, University of Bristol, UK, and Honorary Professor of Hospitality Management at TUT, Taiwan. He is an internationally recognized and awarded scholar specializing in marketing innovation and digital transformation in the service industries. This is evidenced by awarded grants of more than £15 million in total from UK Research and Innovation (UKRI) and the European Union, international invited conference keynote speeches, three international best paper awards and a certificate of achievement for his services as Associate Editor. His research projects focus on decision-making in consumer behaviour as influenced by new technologies, tourism marketing and innovation. Stylos has published in various journals including *Tourism Management, Journal of Travel Research, Psychology & Marketing, Journal of Business Research, International Journal of Contemporary Hospitality Management, International Journal of Hospitality Management* and *Technological Forecasting and Social Change.*

Ellis Urquhart is Lecturer and Postgraduate Programme Leader in Tourism Management within the Business School at Edinburgh Napier University, UK. He specializes in visitor attraction management, co-creative experience design and technological mediation in the tourism and heritage sector. His PhD thesis explored the role of interactive technologies in the co-creation of Scottish visitor attractions. Ellis teaches tourism management at both undergraduate and postgraduate levels at Edinburgh Napier University in addition to overseas programmes delivered in Switzerland, Singapore, Macau and Hong Kong. He currently reviews for a range of tourism publications, has appeared at a number of internationally recognized conferences and contributes to various academic publications.

Chris A. Vassiliadis constitutes member of the academic teaching and research staff of the Department of Business Administration of the University Macedonia in Thessaloniki, Greece. He is a professor and member of University Macedonia teaching staff with more than 20 years of experience working with Greek and non-Greek Universities at the graduate and postgraduate levels. Specialities include management, marketing, and services management and marketing. His research has been published in various journals. Also, he was a collaborating member of the Greek Open University in the unit of Business Administration and Organizations, and is today a collaborating member of the Open University of Cyprus. He was and is part of the teaching staff of different interdepartmental and master's Programmes like the Postgraduate Study in Business Administration of the University of Macedonia, the International Master Program of Business Administration (IMBA) University of National Chiayi University, Chiayi, Taiwan, Athens University of Economics (MSM-OPA), Master in Services Marketing (MSM) a.o. Since 2017, he has been the Director of the Master's Programme in Tourism Management at the University of Macedonia. He has also participated in various international congresses; he is a reviewer of several academic journals and is the writer and co-writer of the following books: *The Administration and Marketing of Tourism Destinations* (in Greek), *Customer Relationship Management and Marketing of Tourism Destinations* (in Greek), *Principles and Practices of Small-Scale Sport Event Management* and *Event Marketing Management.*

Orhan Can Yılmazdoğan is an assistant professor at Eskişehir Osmangazi University, Tourism Faculty where he teaches courses on marketing, digital marketing and cultural tourism. His research focuses on marketing, social media and consumer behaviour. He has published scientific articles and book chapters in the field of tourism. He has experience in scientific researches and project management for more than 10 years. He is responsible for developing the content of educational courses.

Georgia Zouni is Assistant Professor of Tourism Marketing at the Department of Tourism Studies at the University of Piraeus. Her academic interests focus on destination marketing, destination experience, events marketing, tourism satisfaction and digital marketing.

Jeremy Zwiegelaar is currently the subject coordinator and Senior Lecturer in Entrepreneurship at Oxford Brookes University, UK. He has a PhD in Management (Massey University NZ), MA Hons first class in

Psychology (University of Auckland, NZ) and a Graduate Diploma in Teaching (University of Auckland, NZ). He is the Co-director of the research centre Oxford Research on Innovation and Entrepreneurship Lab in the Oxford Brookes Business School. He researches the area of nascent entrepreneurship with a focus on new venture performance. He focuses on business model innovation, digital transformation and well-being of entrepreneurs. He also studies and considers aspects of performance, success and failure outcomes, and factors affecting or explaining this phenomenon. His work has appeared in *Journal of Small Business Management*, *R and D Management*, *Strategic Change* and *International Journal of Contemporary Hospitality*, among others.

List of Figures

List of Tables

1

Introduction: The Shift in Contemporary Service Marketing Management in Tourism and Hospitality

Nikolaos Stylos, Roya Rahimi, and Peter Robinson

Hospitality and tourism comprise the second largest industry in the world and support job growth and economic prosperity, accounting for 1 in 11 jobs worldwide and delivering 1 in 4 of all new jobs created in 2022, that is, 21.6 million new jobs which represents an 22% increase on a year-to-year basis. It is projected to create more than 130 million additional jobs over the next 10 years, and the industry itself will continue growing, thus supporting national economies to mitigate the repercussions emerging from challenging geopolitical factors (WTTC, 2023).

N. Stylos (✉)
University of Bristol Business School, Bristol, UK
e-mail: n.stylos@bristol.ac.uk

R. Rahimi
University of Wolverhampton Business School, Wolverhampton, UK
e-mail: roya.rahimi@wlv.ac.uk

P. Robinson
Leeds Beckett University, Leeds, UK
e-mail: peter.robinson@leedsbeckett.ac.uk

N. Stylos et al. (eds.), *Contemporary Marketing Management for Tourism and Hospitality*, https://doi.org/10.1007/978-3-031-65049-9_1

The marketing discipline is developing quickly and new marketing streams which are influenced by the latest advances in sustainability and social marketing, and two completely new generations of consumers who were born in the Industry 4.0 era shape the new background and trends (Pekar et al., 2024; Stylos & Vassiliadis, 2023). Moreover, the advent of digital technologies and their penetration into marketing management have created new opportunities and challenges for marketing academics and practitioners (Acikgoz et al., 2023, 2024; Shin et al., 2022; Stylos, 2022).

Over the next two decades, GenZers and Gen Alpha individuals will be the driving consumption force who are exceptional users, consumers and producers of Industry 4.0 applications (Priporas et al., 2017; Stylos et al., 2021). They also have a specific mindset towards sustainability and understand that their future on the planet heavily depends on their production and consumption patterns, that is, behaving responsibly and with respect to the natural environment and the collective interest of the local communities.

The purpose of this book is to focus on the contemporary marketing and management aspect of the tourism and hospitality industry as this will be shaping via Smart innovations, Industry 4.0 apps, priority over sustainable futures, and creativity, to shape consumer/tourist expectations, preferences, and behaviors. The latest technological advancements and the challenges associated to a sustainable planet will be decisive factors in tourism and hospitality marketing, considering that GenZers will be the ones leading and shaping the future of this industry. Consequently, through a collection of 13 chapters, this book offers key insights into numerous aspects of the marketing management of the hospitality and tourism industries from a contemporary viewpoint.

In Chap. 2, Belenioti, Stylos, and Vassiliadis explore how smart apps support the transition from consumption to prosumption by implementing digital transformation at the post-travel level/stage within a cultural tourism context. Then, in Chap. 3, Khan provides an in-depth coverage of social media marketing strategies and tools in the context of hospitality and tourism. Pergelova, Beck, Zwiegelaar, and Stylos discuss the impact of new technologies on individuals and organization in Chap. 4, focusing on how SMEs in the hospitality sector can use smart technologies to enhance customer experiences and interactions to build organizational resilience. In Chap. 5, Urquhart explores the increasing hybridization of the servicescape in tourism and hospitality environments, considering the nexus between physical, ambient, and digital resources within the experiential space. Furthermore, Yılmazdoğan, Ece, and Ezenel offer in Chap. 6 an in-depth account of the development of new technologies and their implications for tourism with respect to studying their contribution to creativity and marketing within the process of digitization in the tourism industry.

In Chap. 7, Zouni, Katsanakis, and Klintsaris examine the tourism behavior of Millennials and GenZers, the impact of technology on their tourism behavior, as well as the impact of the COVID-19 healthcare crisis on their travel behavior. Then, Chap. 8, Pantano, Scarpi, and Hu critically discuss the travel behavior of Baby Boomers, Generation X, Y, and Z with respect health risks and the influence of risk perceptions on their decision-making and destination image formation. Moving on to Chap. 9, Yılmazdoğan, Ece, and Ezenel examine the role GenZers play as marketing managers, who are or will be working in the tourism sector in the future. In Chap. 10, Li and Agyeiwaah have put together about Chinese' GenZ perspective on restoring destination image with respect to unsafe places in the post COVID- 19 period. Chapter 11 presents Maton's work on analyzing how responsible tourism marketing and management can be crucial in achieving the United Nations Sustainable Development Goals (SDGs). Chapter 12, authored by Gkarane and Vassiliadis, delves into fresh perspectives on coastal area tourism destination development, innovative underwater practices and technology management to enriching the experience of underwater life. Finally, in Chap. 13, Balan and Escobar explore a new concept of tourism and detach from the strategies pursued by major destinations and rather highlighting natural and cultural heritage elements for the marketing and management of local destinations.

References

Acikgoz, F., Perez-Vega, R., Okumus, F., & Stylos, N. (2023). Consumer engagement with AI-powered voice assistants: A behavioral reasoning perspective. *Psychology & Marketing, 40*(11), 2226–2243.

Acikgoz, F., Stylos, N., & Lythreatis, S. (2024). Identifying capabilities and constraints in utilizing blockchain technology in hospitality and tourism. *International Journal of Contemporary Hospitality Management*. https://doi.org/10.1108/IJCHM-07-2023-1083

Pekar, V., Candi, M., Beltagui, A., Stylos, N., & Liu, W. (2024). Explainable text-based features in predictive models of crowdfunding campaigns. *Annals of Operations Research*, 1–31.

Priporas, C. V., Stylos, N., & Fotiadis, A. K. (2017). Generation Z consumers' expectations of interactions in smart retailing: A future agenda. *Computers in Human Behavior, 77*, 374–381.

Shin, D., Stylos, N., Asim, M., Park, Y., & Williams, R. B. (2022). How does the blockchain find its way in the UAE? The blockchain as a sociotechnical system. *International Journal of Technology Management, 90*(1–2), 122–140.

Stylos, N. (2022). An integrated duality theory framework (IDTF): Marking pathways for consumer decision-making researchers in the hospitality and tourism industry. *International Journal of Contemporary Hospitality Management, 34*(7), 2597–2619.

Stylos, N., Rahimi, R., Okumus, B., & Williams, S. (2021). *Generation Z marketing and management in tourism and hospitality*. Springer International Publishing.

Stylos, N., & Vassiliadis, C. A. (2023). Gamification design: Toward developing image perception scales for generation Z consumers. *International Journal of Contemporary Hospitality Management*. https://doi.org/10.1108/IJCHM-12-2022-1616

WTTC. (2023). *Economic impact research (EIR)*. Retrieved October 24, 2023, from wttc.org

2

From Consumption to Prosumption: The Interplay of Digital Transformation, Prosumers and Post-tourism Experience Within Cultural Tourism

Zoe Charis Belenioti, Nikolaos Stylos, and Chris A. Vassiliadis

Introduction

Recent evidence suggests that major advances in the Web have led to multidimensional positive unprecedented changes at societal, economic, and business levels. Digital transformation (DT) has been implemented through various advancements, including citizen communication with

Z. C. Belenioti (✉)
York St John University, London Campus, London, UK
e-mail: zoibelenioti@gmail.com

N. Stylos
University of Bristol Business School, Bristol, UK
e-mail: n.stylos@bristol.ac.uk

C. A. Vassiliadis
University of Macedonia, Thessaloniki, Greece
e-mail: chris@uom.edu.gr

N. Stylos et al. (eds.), *Contemporary Marketing Management for Tourism and Hospitality*, https://doi.org/10.1007/978-3-031-65049-9_2

diverse media channels and brands, which enables brand co-creation and the producer/prosumer era. This transformation also involves the digital shift in the customer journey, characterized by an increased number of touchpoints and the emergence of the digital customer experience (CX). From B2b to B2c level, from FPOs to NPOs, from apparel to tourism industry, the prosumption phenomenon has become a central issue for marketing and customer experience at pre-purchase, purchase, and post-purchase levels. Likewise, in the context of cultural products, Web 2.0 and interoperability facilitate visitors to become "prod-users".

Within cultural tourism and digital cultural heritage, the toolbox of digital transformation (mobile apps, VR, AR, Big Data, social media) gives a fertile and joyful context for brand co-creation experiences. However, within cultural tourism despite its potentiality, the notion of prosumption at the post-tourism level is still in its infancy. Based on literature review analysis and multiple case analyses from museums, cultural tourism, and the digital cultural heritage industry, this chapter attempts to remedy this gap by exploring for the first time how DT shapes the prosumption phenomenon at the post-tourism level in the cultural tourism sector. This work significantly contributes both to academia and to industry by setting out the groundwork for future research into this new infant stream and by setting out new future research directions and practical recommendations.

Value Co-creation Definition, Historical Perspective, Prerequisites, and Types

Over the past decade, small-scale research began to emerge setting the foundation of value co creation. According to Bal (2019), Prahalad and Ramaswamy were the first to coin the term "value co-creation" referring to it as a collaborative process for creating business value with, rather than just for, customers. Some years later, Ramaswamy accentuated the beneficial role of this experiential, reciprocal collaboration, and VCC for sustaining differentiation and competitive advantages (Bal, 2019). Okoeguale (2023) exploring the major drivers for brand value co-creation identified the following certain reasons: (1) innovation; (2) resource integration; (3) relationship building; (4) collaboration; (5) customer satisfaction; (6) brand equity.

Moreover, as Bal (2019) reported in his thorough review, a considerable amount of literature has been published on the dynamics between stakeholders and VCC process, delineating the diverse and disciplinary nature of VCC. More particularly, according to Bal (2019), the moderators of the symbiosis between stakeholders and VCC may be categorized into the following 11 parameters (1) Economic Factors, (2) Experiential (symbolic and emotional) factors; (3) Societal factors, (4) Relational factors, (5) Functional factors, (6) Characteristics of the platform (trust, transparency, accessibility, flexibility, adaptability, effectiveness, responsiveness, participation, mutual respect/commitment, continuous feedback/support and less bureaucratic), (7) Environmental factors (CSR dimensions, societal justice, combining resources, leveraging knowledge, and capabilities), (8) IP factors related to intellectual capital (knowledge of offerings, possible developments, innovation, creativity, and information on market demand, current trends), (9) Long-term strategical factors (possibility of a long-term relationship, competitive advantages), (10) Factors related to leadership, and (11) Power and interest analysis.

Furthermore, more recent attention has focused on the participatory and active role of consumers (Yi & Gong, 2013; France et al., 2018; Merz et al., 2018; Bal, 2019; Casais, 2020; Darmayanti, 2023). Casais (2020) points out that the essence of value co-creation lies in the collaborative, dialogic, and interactive dynamic between customers, suppliers, and stakeholders and production, the horizontal character of value co-creation harnessed in any business market, both business-to-business and business-to-consumer. "In parallel, value co-creation also arises from customer-dominant logic, where customer-to-customer interaction (CCI) is a driver for customer satisfaction" (2020, p. 490). In the same vein, Darmayanti (2023) denotes that the emergence of value co-creation entails to the new era of marketing and the "shift from goods-dominant (G-D) logic to service-dominant (S-D) logic" (2023, p. 1). Moreover, both France et al. (2018) and Darmayanti (2023) acknowledge the tremendous contribution of customer brand co-creation behaviors, given that the customer becomes the moderator factor as being the influential creator of the whole brand experience and "their voice is influential in brand management" (Darmayanti, 2023, p. 3).

Hence, according to Merz et al. (2018) and France et al. (2018), brand co-creation behavior is a broader term expanding the boundaries of purchase and consumption stages of customer experience funnel. On the contrary, brand co-creation is all about the voluntary active positive attitude customers choose to show toward brand. This positive way of brand experience is characterized "by vigor, dedication, and absorption toward brand co-creation" (France et al., 2018, p. 336). Thus, as Merz et al. (2018) observe, connectedness, creativity, and customer motivations such as passion, trustworthiness, and commitment are essential elements that drive brand co-creation. These elements facilitate the relationship between brands and consumers, leading customers to become prosumers through various types of brand co-creation behaviors such as feedback behavior, advocacy behavior, and helping behavior, as those are identified by Yi and Gong (2013). Additionally, recent studies clearly (Nkoulou Mvondo et al., 2022; Okoeguale, 2023) indicate that that there is a relationship between customer satisfaction and brand value co-creation and that co-creation experience dimensions positively impact brand evangelism, trust, and passion.

However, despite the importance and potential of brand co-creation, very little is currently known about customers role in brand management both at academic and industrial levels. Belenioti has repeatedly called and welcomed new studies within the interplay of various clusters social media users and brand co-creation outcomes not only in for profit organizations (Belenioti, 2015, 2020), but also more importantly within non-profit organizations, cultural tourism products, and cultural institutions (Z. Belenioti & Vassiliadis, 2016; Vassiliadis & Belenioti, 2017; Belenioti, 2022). Similarly, according to Darmayanti (2023), many firms either don't know about the role of their customer to brand management or even if they the firms are aware about the active role of consumers in the brand co-creation, they don't know how to handle different buyer personas and the consequent different motivations of brand co-creation. Furthermore, up to now little evidence has been found associating brand co-creation in cultural institutions with digital transformation and digital marketing. This chapter attempts to remedy this gap.

Defining Terms and Setting the Historical Perspective of DT

Over the past decade, the rise of the Web 2.0 along with new digital technologies and the infant stage of digitization massively transformed the previous notion either the nature of business model and business value chain and the related managerial or policy implications or the nature of business and personal communication (Bounfour, 2016; Schallmo & Williams, 2018; Vial, 2019; Belenioti, 2021; Nambisan et al., 2019). Moreover, in the dawn of DT and the rise of Metaverse, vast changes were enabled in the new digital and social media ecosystem in terms of privacy, dynamics between brand and consumers, presumption and H2H communication (Assiouras et al., 2019; Del Chiappa, 2022).

According to Vial (2019), the set of these novel technologies has a disruptive impact at three major levels: (1) consumer behavior and expectations, (2) disruption of competitive landscape and value chain, and (3) boost of the availability of data (Stylos et al., 2021a, b). Particularly, when it comes to the shift of consumer behavior, the widespread adoption of these digital media, used for both social and business reasons, led to a brand-new communication landscape and users' patterns (Belenioti, 2015, 2017; Belenioti et al., 2015; Vassiliadis & Belenioti, 2017; Dey et al., 2020; Darmayanti, 2023; Rahimi & Kozak, 2017; Rahimi & Gunlu, 2016; Koseoglu et al., 2016). As noted by Dey et al. (2020), digital consumer empowerment is a new reality of Web 2.0 and new Digital Era. Thanks to the interactive communication and participatory culture of Web 2.0, consumers have the power to be heard, to gain critical mass, and, of course, to influence public sphere and moderate in a positive or in a negative way brand management setting the premises of the brand co-creation new experiential marketing era.

The milestone of this transformation lies in the advent of social media (Belenioti, 2020). As Belenioti et al. (2015) note, "social media applications are becoming the new communication status quo". Simultaneously, social media is becoming fast a key instrument and lucrative field from the business perspective thanks to the high consumer engagement offered through social media and web interactivity. Not only do social media constitute a cost-effective, interactive, and, above all, targeted communication tool for every organization, but they also create a co-value for any product or service that can be communicated. Domino's Pizza exemplifies the co-creation product: the firm invites their customers to create their own pizza named "Dominos Pizza Talent" through a contest within their corporate Facebook channel (Belenioti et al., 2015, p. 176). In the same vein, Vial (2019) discussed the case of DBS Bank, which is actively pursuing a full transition to mobile digital banking. This example underscores the importance of co-creation and customer involvement in the new digital era.

Similarly, the rise of Future Internet and Digital Transformation, along with the enabling effects of decentralization is reshaping the value chain. The new state of digital toolkit disrupts the established notion of business operation, products, and barriers to entry by generating new digital products (Apple, Spotify, Netflix) or by rebutting the notion of certain domain of creative industries (Schwertner, 2017; Schallmo & Williams, 2018; Mergel et al., 2019; Nambisan et al., 2019; Vial, 2019; Verhoef et al., 2021) or even cultural industries (Belenioti & Vassiliadis, 2016; Belenioti et al., 2017a, b; Vassiliadis & Belenioti, 2017; Belenioti et al., 2019a, b; Agostino et al., 2020; Gannon et al., 2022; Bihamta et al., 2017; Rezaei et al., 2018; Sigala et al., 2019; Rahimi et al., 2023; Belenioti, 2021; Raimo et al., 2021; Simone et al., 2021). All in all, Future Internet transforms indeed the business value not only by moving online any business operation, but more importantly by setting boosting the brand co-creation experience (Assiouras et al., 2019; Del Chiappa, 2022). This shift is also evident not only within creative and cultural industries but also within the Healthcare sector, as brand co-creation is the new locus on patient experience. Reciprocity between patients and their healthcare professionals, and customer-visitor-patient empowerment, is the new rule in the game called patient experience, as doctors don't have the power and patient brand value co-creation is the moderator factor of this dynamic (Darmayanti, 2023).

Parallelly, digital technologies also boost the generation of data and data mining via the retrieval either of cookies or of the user patterns within mobile devices, social media ecosystem, user attitudes, and reaction toward enterprise social media accounts or even by using data-driven algorithmic decision-making (Tommasetti et al., 2010; Buhalis & Inversini, 2014; Vassiliadis & Belenioti, 2017; Cerquetti & Ferrara, 2018; Dey et al., 2020; Martins et al., 2020; Pantano & Stylos, 2020; Del Chiappa, 2022). Via social media and digital marketing toolbox not only do firms may have an additional operational channel but they can also monitor their brand performance (Buhalis & Inversini, 2014; Vassiliadis & Belenioti, 2017; Vial, 2019; Raimo et al., 2021; Simone et al., 2021; Belenioti, 2021).

However, according to Schallmo and Williams (2018), despite the surge of interest on digital transformation since 2016, the notion of digital shift of products and services emerged in the 1990s and 2000s, via the advent of mass media and web2.0—smart connectivity and smart devices, respectively. The first major milestone was the advent of mass media communication, which facilitated mass marketing throughout the 1990s and 2000s. Next, from 2000 to 2015 the second milestone led to the customization, personalization, brick and mortar, omnichannel and real time and reciprocal without borders digital communication and transaction. Thus, as Schallmo and Williams (2018) emphasized, before delving into the term "digital transformation", it is important to understand exactly what/how the term "digital" is defined in digital transformation and what are the certain stages of digital transformation. According to Schallmo and Williams (2018), "McKinsey which states that digital is less about any one process and more about how companies run their business with the aim of (a) Creating value at the new frontiers of the business world (b) Optimizing the processes that directly affect the customer experience, (c) Building foundational capabilities that support the entire overall business initiative" (2018, p. 3).

Furthermore, at this point the clarification regarding certain stages of digital transformation and the difference between the terms "digitization", "digitalization", and "digital transformation" is deemed useful. So far, the terms "digitization" and "digital transformation" are used interchangeably without a clear distinction. In a nutshell, digitalization is the switch from analog to digital or how IT or digital technologies are harnessed to change or improve focusing on cost savings, or on the enhancement of customer experience. On the other hand, the European Commission defines digital transformation as the integration of digital technologies by business, which also includes the societal impact of innovations like the Internet of Things (IoT), cloud computing, digital platforms, and blockchain technologies, (Belenioti, 2022, p. 409).

Phases and Key Success Factors in Digital Transformation

Verhoef et al. (2021) identify "three stages for digital transformation: digitization, digitalization and digital transformation. Each phase places specific demands on firms' digital resources, organization structure, growth strategies and metrics" (2021, pp. 890, 895). Verhoef et al. (2021), agreeing with Schallmo and Williams (2018), define digitization as "the action to convert analog information into digital information. Examples concern the use of digital forms in ordering processes, the use of digital surveys, or the use digital applications for internal financial declarations. Typically, digitization mainly digitalizes internal and external documentation processes, but does not change value creation activities" (2021, p. 891).

On the contrary, consistent again with Schallmo and Williams (2018), Verhoef et al. (2021) accentuate that digitalization describes "how IT or digital technologies can be used to alter existing business processes. In digitalization, IT serves as a key enabler to seize new business possibilities by changing existing business processes, such as communication, distribution, or business relationship management (2021, p. 891). In other words, any change or improvement that firms make by leveraging new digital technologies to enhance the user or customer experience—such as through mobile or online communication channels—constitutes digitalization. Successful digitalization requires tech-savvy employees and promotes a positive organizational culture towards embracing technology.

Moving to the digital transformation definition, as Belenioti (2022, p. 409) notes, the European Union defines digital transformation as the "integration of digital technologies by European enterprises and the impact on society of new technologies, such as the Internet of Things

(IoT), cloud computing, innovative digital platforms and blockchain technologies" (2022, p. 409). The ultimate goal of digital transformation is to create disruptive cutting-edge business models and new product or that services improve the quality of citizens lives by enabling high quality, real time, personalized and interactive services (Nambisan et al., 2019; Mergel et al., 2019; Schwertner, 2017). According to Verhoef et al. (2021), digital transformation is the most "pervasive phase, and describes a company-wide change that leads to the development of new business models" (2021, p. 891). Hence, IT here is not a catalyst to optimization but a driver of new business operation and source so of a new competitive advantage based on profound changes to existing business routines. Technology as the key to transform the whole firm will leverage the core competencies or create new, like in the case of healthcare, tourism, and education.

Verhoef et al. (2021) identified two major external factors of digital transformation. Firstly, the widespread adoption of the Web creates a brand prominent new digital market. Secondly, the rise of e-commerce that gives a birth to various niche market and products of accompanying technologies (e.g., broadband internet, smartphones, Web 2.0, SEO, cloud computing, speech recognition, online payment systems, and crypto currencies). Thirdly, the rival completion due to the technological advance within FPOs and NPOs context, and finally the consequent new behavioral patterns accelerated the digital transformation.

In a nutshell, the successful business digital transformation depends on a chain of several factors and dimensions (Kane et al., 2015; Bounfour, 2016; Schwertner, 2017; Vial, 2019; Verhoef et al., 2021). As Schwertner (2017) report, within digital business transformation, big data initiatives and investment on analytics, cloud, mobile, and social technologies may serve as moderator factors and assets of the infrastructure, as these technologies giving the first mover advantages will lead to bigger revenues and a bigger market share in line with the leaders' patterns. Additionally, according to Schwertner (2017), successful digital transformation prerequires the transformation of organization culture and organization structure along with the strategic integration of the digital technologies. Thus, digital skills and leadership, organization structure, culture, and digital transformation strategy should be the essential milestones of the digital transformation roadmap. However, he also alerts on the barriers of DT such as data security issues, lack of interoperability with existing IT systems, and lack of control.

Bounfour (2016), agreeing with Bharadwaj et al., emphasizes four dimensions: (1) the scope/extent of DT, (2) the scale defining the emergence and abundance of platforms that create powerful network leveraging data, (3) the speed, which defines acceleration and in terms of launching products/services, decision making, building networks and finally the (4) source of value creation (data, networks, digital architecture). Similarly, according to Schwertner (2017), Bounfour (2016), exploring the success factors of DT, recognizes the importance not only of corporate digital strategy but also for the governance structures, digital policy, and leadership skills. In the same vein, according to Nambisan et al. (2019), openness of innovation (nature and extent that DT facilitates in the dual terms of actors and contributors, affordances (action or potentiality of action) and generativity (capacity performed by digital technologies to cause disruption via "blending" or recombination are the three key components that define the success of DT.

Furthermore, Kane (2015), clarifying certain misconceptions and myths on DT, supports that DT does not fit to all organizations, and point outs that "disruptive" impact of these digital innovations derives more on from "conventional" operational strategic technology rather than the technology itself. Moreover, doubting that most executives want to transform big brand and market leaders, as small companies are more positive to DT. On the question of DT barriers within higher education (HE), Gkrimpizi and Peristeras (2022) identified that (a) contextual, (b) social, (c) technical, and (d) cultural barriers impede DT in HE organizations. Additionally, Suárez et al. (2021) accentuated the alignment between DT strategy and organizations' vision and acknowledged the beneficial contribution of DT on management resources.

Defining Cultural Institutions, Cultural Products and Industries, Cultural Tourism, Culture, Cultural Goods, and Cultural Heritage

Before proceeding to examine the link between DT and Prosumption within Cultural Tourism, it is important to define the terms "culture", "cultural goods and industries", "cultural tourism", and "cultural heritage". In 2009 UNESCO (2009) used the term "culture cycle" to describe the interchange and interconnections among cultural products activities and the resultant feedback processes by which activities of cultural consumption inspire the creation of new cultural products and artifacts. "Cultural goods convey ideas, symbols, and ways of life, some of which may be subject to copyrights. Cultural activities embody or convey cultural expressions, irrespective of the commercial value they may have. These activities may be an end in themselves, or they may contribute to the production of cultural goods and services (UNESCO-UIS, 2005)" (2009, p. 23).

Aiello and Cacia (2014) define culture as "common good, a key element of community's wellbeing, for the entire humanity (refer to the definition and classification of Cultural Heritage adopted by UNESCO), and as such everyone is responsible for preserving, maintaining, and developing it". For Aiello and Cacia (2014), culture tends to be broadly used to refer to the any common cultural product (either it is good, service or idea) that contributes to exchange of value for good, consistent and shared economic, social and cultural models, always in line with the UNESCO's definition of World Cultural Heritage for global access of any citizens to culture. As Aiello and Cacia (2014) assert, "Everyone must be enabled to enjoy culture"—"Culture is a world heritage site" (UNESCO, 2005) (2014, p. 1).

In terms of the breadth and classification of cultural sector, UNESCO (2009) reports that although there is a consensus that culture is an umbrella term of on interconnecting activities, still there is an ambiguity on the classification of the cultural activities. Thus, so far cultural sector encompasses the following six domains belonging to the Intangible Cultural Heritage (transversal domain): (1) Cultural and Natural

Heritage; (2) Performance and Celebration; (3) Visual Arts and Crafts; (4) Books and Press; (5) Audio-visual and Interactive Media; and (6) Design and Creative Services. Beyond the six aforementioned cultural sector domains, Tourism and Sports and Recreation are related domains to the cultural cycle. Furthermore, UNESCO concludes that the following three additional transversal domains of (a) Education and training, (b) Archiving and preservation and (c) Equipment and supporting materials massively contribute to the culture cycle for the production, transmission of culture.

Cultural goods belong to cultural industries. For Aiello and Cacia (2014), cultural industries is an umbrella definition of commodified activities, ranging from the mass media (print and broadcast media and advertising), film, new media, to art, design, music, and architecture. However, despite the contribution of cultural industries to economy and society, there is a paucity of research regarding cultural industries as Aiello and Cacia (2014) based on relevant insights (Rahimi, 2017; Rahimi et al., 2017; Vatankhah et al., 2023). Similarly, Riche Resources define cultural institutions as "institutions with an acknowledged mission to engage in the conservation, interpretation and dissemination of cultural, scientific, and environmental knowledge, and promote activities meant to inform and educate citizens on associated aspects of culture, history, science and the environment" (https://resources.riches-project.eu/glossary/cultural-institutions/). Cultural institutions comprise (1) museums, (2) libraries, (3) historical or botanical societies, and (4) community cultural centers (Riche Resources, 2014). Additionally, as noted by Riche Resources, the ultimate goal of cultural Institutions is to boost citizens' interaction and engagement with heritage building bridges via cultural understanding, intercultural dialogue, and cultural diversity. In a nutshell, cultural institutions facilitate the conservation, revitalization, interpretation, and documentation of Cultural heritage, as well as the transmission of culture across generations (Riche Resources, EU, 2014).

But what is meant as "cultural heritage"? UNESCO (2014) defines cultural heritage as "both a product and a process, which provides societies with a wealth of resources that are inherited from the past, created in the present and bestowed for the benefit of future generations" (2014, p. 3). According to UNESCO (2014), cultural heritage is classified into

(a) monuments and (b) groups of buildings with homogeneity or their place in the landscape, in terms of architecture with outstanding value from the historical, artistic or scientific perspective; (c) sites: works of man or the combined works of nature and man, and areas including archaeological sites, which are of outstanding value from the historical, aesthetic, ethnological or anthropological point of view" (2014, p. 4). The term "cultural heritage" encompasses several main categories of heritage such as (a) cultural heritage including tangible and intangible cultural heritage resources, including movable, immovable, and underwater cultural heritage; (b) Natural heritage, including natural sites with cultural aspects such as cultural landscapes, physical, biological, or geological formations; and (c) heritage in the event of armed conflict (UNESCO, 2009).

Besides its beneficial contribution to future generation mental and cultural enrichment, there is also abundant evidence that cultural heritage, or its resultant cultural products such as cultural institutions (museum or cultural centers) or certain cultural tourism dimensions as defined by Ohridska-olson (2015) as festival or event cultural tourism, arts-defined cultural tourism, creative tourism or Urban Cultural Tourism), plays a crucial role in boosting the understanding cultural diversity. Historically, cultural heritage has been served as a vehicle and drive of social cohesion by either building bridges towards a more cohesive and open society as a window to the world, or by formulating common grounds within cultural heritage resources and connecting people on the basis either of the common interest of the cultural products (EU, 2014; Hesmondhalgh & Pratt, 2005; UNESCO, 2009; Aiello & Cacia, 2014).

Particularly, as the European Commission highlights especially during the current tough times of multi-crisis such as the increased migration toward several Member States, the increased mobility of people across the EU and the emerging climate changes and sustainability issues, "cultural diversity is an asset, an added value and an opportunity for European societies has permeated the work of our expert group, convened in the framework of the Agenda for culture to focus on the role of public arts and cultural institutions in the promotion of cultural diversity and intercultural dialogue. (...) Cultural institutions need to engage broader parts of the population to serve their core mission of ensuring the transfer of cultural heritage across generations. (…) Doing so might require adopting new methods of working which allow for innovation and room for experimentation, while maintaining excellence and high artistic and cultural standards offered to audiences and participants" (EU, 2014, p. 7).

Cultural Heritage in the New Digital Era

Undoubtedly, the rise of Internet along with Web 2.0 has introduced a brand-new era of the digital cultural heritage particularly reflected on certain cultural institution and industries such as art organizations and museums. Museums are competing with various cultural institutions and entertainment facilities (Fletcher & Lee, 2012; Hausmann, 2012a, b; Kim & Ko, 2012).

As Belenioti states, "The positive prospects of this market, along with increasing competition and technological advances have totally changed the nature of museums introducing a new museum profile and experience. This new profile of museums has two main components: the emergence of cultural marketing and the impact of social media" (Belenioti, 2020, p. 21). Thus, museums are adapting to the new digital era by implementing suitable digital tools to engage and attract online audience (Kotler, 2001; Hume & Mills, 2011; Colbert & Courchesne, 2012). More and more museums are increasingly adopting the digitalization and personalization coming from Web 2.0. Undoubtedly, digitalization expanding the relationship between museums and visitors enables the embryonic stage of digital transformations. Consequently, visitors seek information anytime and anywhere. Moreover, Lepouras and Vassilakis (2004) argue that Web 2.0 leads museums to the new stream of edutainment (education+entertainment provided by combination of museum and web tools). This new stream boosts visitors' engagement in the offline environment and it provides new educational opportunities (Marty, 2011). Typical examples of edutainment constitute personal digital collections such as Getty Museums at Tate, Educators Online at the Museum of Fine Arts, Boston, Learning@Whitney at Whitney Museum of American Art, and RFID application at the Science Museum in London.

Simultaneously, "social media applications are becoming the new communication status quo". Social media offer dialogic, well-targeted, and cost-effective communication. Social media instruments have extended the notion of interaction, while they also provide visitors with new chances of interaction beyond the offline museum via the 3D museum representation or artifacts such as social media and video, e-database, and digital museum collections (Lepouras & Vassilakis, 2004; Marty, 2008; Arends et al., 2009; Weilenmann et al., 2013). However, social media pose crucial challenges for museum managers. To begin with, museums fall behind in creating dialogic communication via social media. Similarly, Quinton and Fennemore (2013) observe that though NPOs have integrated Web 2.0 tools, they have adopted one-way communication, totally different to the e-marketing orientation. Furthermore, museums find attracting and retaining visitors' loyalty difficult, although visitors stand at the center of the museum experience. As a result, effective communication policy is one of the greatest challenges for museums.

Overall, the pervasiveness of social media within the social and business context along with the new digital face of museums has attracted the attention of museum scholars and practitioners. Online marketing tools and social media appear especially effective in arts institutions. Colbert and Courchesne (2012) emphasize the positive contribution of social media instruments to the performance of cultural industries. In this vein, recent evidence shows that cultural heritage products seem to be benefited by the use of the state of the art technology (Mobile app, QR codes, Social Media) launching a whole new interactive and highly prominent user experience (Bakhshi & Throsby, 2012; Aiello, 2014; Corte, 2014; Palumbo et al., 2014; Tardivo et al., 2014).

The encouraging prospects of this market, the increasing competition, and the emergence of Web 2.0 have created new challenges and opportunities for successful marketing strategies within cultural organizations (Schweibenz, 1998; Froeschl et al., 2008). Furthermore, Belenioti, conducting a series of studies on the digitalization within museums and cultural institutions (Belenioti & Vassiliadis, 2016; Belenioti et al., 2017a, b; Vassiliadis & Belenioti, 2017; Belenioti et al., 2019a, b; Belenioti, 2020, 2021), in line with Kim and Ko (2012), concludes that nowadays museums facing fierce competition from various other cultural institutions and entertainment facilities strive for a more efficient, direct, and effective communication policy to ensure their financial viability and sustainability. However, until recently the communication policy of museums is affiliated with more conventional communication tools (Schweibenz, 2004). Due to the lack of effective marketing strategy and proper technological infrastructures, many museums have difficulty attracting and retaining customer loyalty. Moreover, according to Aiello and Cacia (2014), upon the digital convergence and the universal access to cultural products across borders despite WTO guidelines, still certain questions on the nature of culture (good/service) and how cultural products are to be treated in the digital era remain unclear.

Digital Transformation and Brand Co-creation in Tourism and Cultural Products/Institutions

All in all, both the advent of Internet and most importantly the launch of digital transformation using Web 2.0 and Web 3.0 tools such as social media or Future Internet and Internet of Things (IoT) enable the brand co-creation era (Hu et al., 2023). More specifically, under the umbrella of "digital transformation", tourist and museum visitors are now becoming "prod-users".

Using Future Internet and Smart Tourism applications within museums (RFID, Augmented Reality (henceforth AR), Virtual Reality (henceforth VR), Museum Kiosks) visitor communication cues increase brand engagement, museum satisfaction, brand equity enabling the co-creation. Moreover, digital transformation is a moderator factor of the Customer Experience era (Lemon & Verhoef, 2016; Buhalis, 2022) including tourism industry. Thus, according to Lemon and Verhoef (2016), marketing scholars and practitioners should elaborate on the integral role of IT, and how new digital transformation tools such as robots, IoT sensors, and mobile applications could optimize the customer service and boost visitor and customer engagement by delivering a unique memorable customer experience.

Particularly, according to Buhalis (2020), the integration of hardware, sensors, data storage, big data process, and analytics within the promising Ambient Intelligence delivers a new holistic experience. Interconnectivity and interoperability of smart technologies that provide a direct, dynamic, and ongoing personalization by synchronizing personal data and visitors' behavioral patterns is just a preview of the new facet of ICT, Future Internet harnessed by Internet of Things, Smart Tourism Applications, and Big Data (henceforth BD). According to Stylos and Buhalis, BD, intranet reviews and eWOM Metadata, Web and text analytics, data mining and information retrieval based on the online consumers activity may optimize the targeting process by tracking better the tourist behavioral patterns.

Indeed, the widespread adoption of ICT, Web 2.0, and social media (henceforth SM), along with the emergence of smart tourism technologies, was the catalyst for unveiling the new museums experience followed by the digital transformation in tourism. Similarly, in the context of cultural tourism, including museums and festivals, digital transformation—enabled by ICT, Web 2.0, social media, and big data—has redefined the relationship between cultural institutions and their visitors, moving from a traditional business-to-consumer model to a consumer-to-consumer, or human-to-human, communication approach (Belenioti, 2020; Buhalis, 2020). Moreover, according to Raimo et al. (2021), all these new participatory digital technologies have tremendously boosted the promotion of culture, enrichment of cultural on-site visits, and managerial and organizational activities.

According to the recent literature, the milestone of this transformation in cultural tourism lies in the advent of social media (Belenioti & Vassiliadis, 2016; Belenioti, 2017, 2020; Vassiliadis & Belenioti, 2017) thanks to its continuous, dialogic, well-targeted, and cost-effective communication (Belenioti, 2020; Belenioti & Vassiliadis, 2017; Amanatidis et al., 2019; Belenioti et al., 2019a, b). Particularly, social media has a positive contribution not only in terms of enhancement of museums' learning process (Vassiliadis & Belenioti, 2017) but also on brand equity, visitor satisfaction, visitor engagement, museum experience, and the intention to revisit (Belenioti et al., 2019a, b; Belenioti, 2020). Likewise, when it comes to festivals social media have been repeatedly emerged as an efficient awareness and engagement tool (Hudson & Hudson, 2013; Hudson et al., 2015; Hoksbergen & Insch, 2016; Belenioti et al., 2017b; Celuch, 2021) of digital transformation.

In addition, a strong relationship between COVID-19 and digital transformation in museums has been reported in the literature, according to Sigala (2020) and Raimo et al. (2021). Indeed, Raimo et al. (2021), based on multiple case studies in Italian museum context, found that the COVID-19 pandemic has accelerated the adoption of digital technologies. Indeed, several reports have shown that museums globally harnessed new digital transformation tools to stay connected with their museum audience and friends (Burke et al., 2020; Calvert et al., 2020; Course & Systems, 2020; Gretzel et al., 2020; NEMO, 2020; Ou, 2020; Association, 2021).

Brand co-creation, digital marketing, and digital transformation appear to be closely linked. According to Belenioti (2022), in line with Buhalis (2020), Stylos et al. (2021a), and Belenioti (2020), not only did digital transformation via Web 2.0, social media, and big data reshape the relationship between museums and visitors enabling co-creation and H2H communication, but it also boosted engagement, awareness, and expand resilience amid COVID-19 (Assiouras et al., 2019; Bal, 2019; Buhalis & Sinarta, 2019; Belenioti, 2022; Del Chiappa, 2022; Dwivedi et al., 2022).

Indeed, Belenioti (2022), reviewing the link between DT and cultural institutions, concluded that DT tools including social media serve as a first-aid kit and resilience tool in order that museums keep the society alive by moving online the museum/festival experience during and after the pandemic. Moreover, her study confirmed the multidimensional benefits of social media and its role in brand co-creation, supporting earlier research on how social media enhances the flow of museum experiences, facilitates co-creation, and promotes the 'museum beyond the museum' concept (Vassiliadis & Belenioti, 2017; Belenioti, 2020; Hausmann, 2012a, b; Fletcher & Lee, 2012). Recent studies indicate that digital transformation, driven by Web 2.0, social media, and big data, has not only reshaped the relationship between museums and visitors—enabling co-creation and H2H communication—but also boosted engagement, awareness, and resilience during the COVID-19 pandemic (Buhalis, 2020; Stylos et. al, 2021a; Belenioti, 2021).

Belenioti (2022) points out the problems and the challenges on accessibility and financial resources attendant to new museum role as knowledge repositories, communicators, and community connectors. COVID-19 resilience toolbox in museums included social media, 5G, livestreaming, tailor-made events based on visitors' demographics and digital analytics, virtual exhibitions, curatorial talks, and viewing rooms. Tours of current collections, art fairs, and touring exhibits and education/remote learning, such as drawing and photography classes/masterclasses, AR, VR (Kahn, 2020; Practice, 2020) which both Thessaloniki International Film Festival has been incorporating since 2018 (Belenioti, 2022), and Australian Museums, and, lastly, the prominent big data (Kahn, 2020; Practice, 2020) were the backbone of "museum out the museum" and "co-creation museum digital era".

In Italy, museums doubled their social media usage and online activity, and customized their content to remain open during the anthropause (Agostino et al., 2020). Finally, according to Belenioti (2022), in the case of Chinese and Australian museums, digital transformation tools served not only as a brand awareness, brand engagement, and survival tool but also as a test bed of CSR campaigns as museums acted as connector with the healthcare industry.

Much of the literature since the mid-1990s emphasizes the contribution of Web 2.0 and social media to brand co-creation both in FPOs and in NPOs with a specialization to cultural and healthcare NPOs (Bruns, 2012; Bal, 2019; Lee, 2019; Belenioti, 2020, 2022; Martins et al., 2020). Particularly, since 2016 a large and growing body of literature has investigated the role of Internet and DT on brand co-creation of cultural industries such as museums and festivals (Yu & Yao, 2023; San Cornelio & Cruz, 2014; Belenioti, 2017, 2020, 2022; Belenioti et al., 2017b; Vassiliadis & Belenioti, 2017; Wang, 2018; Cerquetti, 2018; Belenioti et al., 2019b; Simone et al., 2021; Bushey, 2023), while some studies have accentuated the contribution of DT and brand co-creation to well-being (e.g. Hu et al., 2023), and resilience of museums' audience (Belenioti, 2022; Bushey, 2023). There is a relatively small body of literature that is concerned with the positive impact of smart tools to the immersive museum experience (Tommasetti et al., 2010; Profile, 2013; Mighali et al., 2015; Bran et al., 2020; Hind, 2022). In the same vein, given the infancy, few studies have appraised the role of IoT to museum brand co-creation (Table 2.1) (Cerquetti & Ferrara, 2018; Li et al., 2021).

Overall, based on the latest research, brand co-creation in Cultural Institutions (CI) is facilitated by the tools outlined in Table 2.2.

Table 2.1 Leading studies in Digital Transformation and brand co-creation in cultural institutions, 2016–2022 (Source: Authors)

Smart tools and museum brand co-creation	Tommasetti et al. (2010), Profile (2013), Mighali et al. (2015), Bran et al. (2020), Hind (2022)
DT as a source of brand co-creation and museums	Yu and Yao (2023), San Cornelio and Cruz (2014), Belenioti (2017, 2020, 2022), Vassiliadis and Belenioti (2017), Wang (2018), Cerquetti (2018), Belenioti et al. (2019b), Simone et al. (2021), Bushey (2023)
IoT & Museums brand co-creation	Cerquetti and Ferrara (2018), Li et al. (2021)
Brand co-creation via in museum as a resilience toolkit	Belenioti (2022), Bushey (2023)

Table 2.2 Digital transformation and brand co-creation toolkit (Source: Authors)

Digital transformation and brand co-creation tools in CI
Big Data
Smart Apps
AR artifacts/events
VR artifacts/events
Responsive website
Online screenings/video
Content accessible to the digitally diverse audience (older or disabled people)
Livestreaming
Online ticketing
Chatbots
Social media
Virtual exhibitions / online talks & masterclass
Gamification
Storytelling in social media
Apps

Importance and Future of DT, Key Themes in DT and Brand Co-creation Future Research Agenda

Recent evidence suggests that the multidimensional positive impact of DT at societal, economic, and business levels (Vial, 2019; Schallmo & Williams, 2018; Verhoef et al., 2021; Lemon & Verhoef, 2016). Indeed, digital transformation and emerging innovative business models have a disruptive impact on the established markets, citizen lives, and consumers' behavioral patterns (Lemon & Verhoef, 2016; Schallmo & Williams, 2018). From citizen reciprocal communication with the plurality media channels and brands enabling the brand co-creation and producer/prosumer era, to the digital shift of the customer journey by increasing number of touchpoints or by emerging digital customer experience (CX), digital transformation has been put in action (Lemon & Verhoef, 2016).

Although technological development and the advent of Web 2.0 have totally changed the marketing communication strategy adopted by arts organizations, there is a scarcity of studies on e-marketing and electronic word of mouth communication (henceforth eWOM) in the field of arts and museum marketing. Despite the digital transformation following the rise of Web 2.0 and social media, little is known about how museums use e-marketing tools to boost their communication and branding (Hausmann, 2012a, b). In addition, although several scholars have investigated the use of social media in museums, so far scholars have merely focused on how NPOs and museums exploit social media (Quinton & Fennemore, 2013; Fletcher & Lee, 2012; Hausmann, 2012a, b). Thus far, no research has categorized the major opportunities, usage patterns and challenges or explored the impact of DT in cultural institutions. Given the infancy of the research stream very little is currently known about the impact of DT at the organizational level in terms of operational efficiency, organization performance, ethics, security, and privacy issues, dynamic capabilities, and digital skills and assets such as capabilities related to digital agility, digital networking and big data analytics (Verhoef et al., 2021). No study so far has developed a new framework regarding digital metrics (Verhoef et al., 2021). Little is known about the integration of DT as integral part of business orientation (Schallmo & Williams, 2018) and the impact that DT will have on organizational structure such as creation of new sections such as digital departments.

Likewise, in the context of cultural products, Web 2.0 and interoperability facilitate visitors to become "prod-users" or having remote access to material online, particularly valued during unprecedented black swan events like the COVID-19 pandemic. Digital transformation toolbox enabled cultural institutions, such as museums, to stay alive by offering more innovative, personalized, active experiences, or shifting to experiential notions of "edutainment" and "artertainment".

However, in terms of the link between brand co-creation and DT in cultural institutions, research is still in its infancy upon the embryonic stage of digital transformation and brand co-creation. Thus, a holistic integration of social media/ DT and smart apps including and the problems and the barriers attendant to their integration as a viable strategic tool for curation and marketing orientation could be usefully explored in further research. "A key policy and research priority should be therefore the astonishing symbiosis of processes, personnel and physical amenities with digital transformation toolbox on boosting satisfaction, relax, well-being, implicit learning during the visit and on emerging museums as ideal escapade, in line with Del Chiappa et al." (Belenioti, 2022, p. 6).

Thus, based on prior evidence on the contribution of deploying repertory grid analysis into segmentation (Vassiliadis & Fotiadis, 2008), of blueprinting process in tourism experience (Vassiliadis & Fotiadis, 2016), and of exploring moderator features triggering visitor satisfaction and revisit intention (Fotiadis, 2018; Vassiliadis et al., 2021), future studies should further delve into behavior patterns of visitors. Last, given the latest work by Fotiadis et al. (2020), a further study could assess destination and event choices as moderator factors of tourism involvement in the post-pandemic era in cultural tourism context.

Moreover, according to Belenioti (2021), improving digital literacy skills and harnessing on a larger scale Smart Tourism Application (AR, VR, Mobile Media apps) to boost accessibility by elder and disabled audience and to provide more personalized experience is still an understudied and prominent research stream. "Future research agenda include directions such as (a) mapping the new digital museum experience,

(b) harnessing AR, VR, big data in a holistic, integrated framework to reach higher engagement and personalization, and (c) call for starting the dialogue on the new "museum era out of museum" era" (2021, p. 1).

Recommended Further Reading

Acikgoz, F., Perez-Vega, R., Okumus, F., & Stylos, N. (2023). Consumer engagement with AI-powered voice assistants: A behavioral reasoning perspective. *Psychology & Marketing, 40*(11), 2226–2243.

Acikgoz, F., Stylos, N., & Lythreatis, S. (2024). Identifying capabilities and constraints in utilizing blockchain technology in hospitality and tourism. *International Journal of Contemporary Hospitality Management*. https://doi.org/10.1108/IJCHM-07-2023-1083

Hu, M., Pantano, E., & Stylos, N. (2023). "Home alone" no more: How does the internet of things (IoT) enhance travellers' subjective well-being. *Technological Forecasting and Social Change, 192*. https://doi.org/10.1016/j.techfore.2023.122563

Jiang, Y., & Stylos, N. (2021). Triggers of consumers' enhanced digital engagement and the role of digital technologies in transforming the retail ecosystem during COVID-19 pandemic. *Technological Forecasting and Social Change, 172*. https://doi.org/10.1016/j.techfore.2021.121029

Priporas, C. V., Stylos, N., & Fotiadis, A. K. (2017). Generation Z consumers' expectations of interactions in smart retailing: A future agenda. *Computers in Human Behavior, 77*, 374–381.

Stylos, N. (2022). An integrated duality theory framework (IDTF): Marking pathways for consumer decision-making researchers in the hospitality and tourism industry. *International Journal of Contemporary Hospitality Management, 34*(7), 2597–2619.

Stylos, N., Fotiadis, A. K., Shin, D. D., & Huan, T. C. T. (2021). Beyond smart systems adoption: Enabling diffusion and assimilation of smartness in hospitality. *International Journal of Hospitality Management, 98*. https://doi.org/10.1016/j.ijhm.2021.103042

References

Agostino, D., Arnaboldi, M., & Lampis, A. (2020). Italian state museums during the COVID-19 crisis: From onsite closure to online openness. *Museum Management and Curatorship, 0*(0), 1–11. https://doi.org/10.1080/09647775.2020.1790029

Aiello, L. (2014). Web 2.0 model in cultural organization: Case study peepul. In *Handbook of research on management of cultural products: E-relationship marketing and accessibility perspectives* (pp. 303–320). https://doi.org/10.4018/978-1-4666-5007-7.ch015

Aiello, L., & Cacia, C. (2014). The cultural product: Integration and relational approach. In *Handbook of research on management of cultural products: E-relationship marketing and accessibility perspectives* (pp. 1–21). https://doi.org/10.4018/978-1-4666-5007-7.ch001

Amanatidis, D., et al. (2019, October 24–27). Cultural communication through social media: Exploring Instagram Practices among Greek museums. In *TOURMAN 2019, 3rd international scientific conference "tourism, travel and hospitality at crossroads: The way ahead.*

Arends, M., et al. (2009). Interaction with art museums on the web. In *Proceedings of the IADIS Intl conference WWWInternet* (pp. 117–125) http://www.ec.tuwien.ac.at/~dieter/research/publications/WWWInternet09.pdf

Assiouras, I., et al. (2019). Value co-creation and customer citizenship behavior. *Annals of Tourism Research, 78*. https://doi.org/10.1016/j.annals.2019.102742

Association, W. M. (2021). *WMA's social media best practices: Help your brand during.* https://westmuse.org/articles/wma's-social-media-best-practices-help-your-brand-during-covid-19

Bakhshi, H., & Throsby, D. (2012). New technologies in cultural institutions: Theory, evidence and policy implications. *International Journal of Cultural Policy, 18*(2), 205–222. https://doi.org/10.1080/10286632.2011.587878

Bal, A. S. (2019, November). *Understanding collaborative value co-creation from stakeholder perspective: A literature review.* Master's thesis.

Belenioti, Z. C. (2015). A snapshot of Greek social media users. In *A snapshot of Greek social media users* (pp. 1–15) http://meaningacrossmedia.mcc.ku.dk/proceedings/.

Belenioti, Z. C. (2017). In L. Steckman & M. Andrews (Eds.), *Online around the world_ Greece_Belenioti.pdf.* ABC-CLIO. http://www.abc-clio.com/ABC-CLIOGreenwood/product.aspx?pc=A4435C

Belenioti, Z.-C. (2020). *The contribution of e-marketing to the branding of non-profit organizations (NPOs): The case of Greek museums*. Aristotle University of Thessaloniki. http://ikee.lib.auth.gr/record/319725

Belenioti, Z.-C. (2021, May). *Covid19 resilience via digital cultural heritage; digital life in museums and festivals during the anthropause* (pp. 19–22).

Belenioti, Z. C. (2022). Covid19 resilience via digital cultural heritage; digital life in museums and festivals during the Anthropause. *International Journal Tecnhology Marketing, 640*(May), 19–22. https://doi.org/10.1504/ijtmkt.2022.10049149

Belenioti, Z. C., Andronikidis, A. I., & Vassiliadis, C. (2015). Classifying and profiling social media users: An integrated approach. In *Proceedings of the second European conference on media & mass communication* (pp. 183–204) http://iafor.org/archives/proceedings/EuroMedia/EuroMedia2015_proceedings.pdf

Belenioti, Z. C., Gkarane, S., & Vassiliadis, C. (2017b). The contribution of social media to small scale events: Boosting the city branding& tourism demand through Thessaloniki Food Festival. In *5th international conference on contemporary marketing issues ICCMI, June 21–23, 2017 Thessaloniki, Greece* (p. 7877).

Belenioti, Z. C., Tsourvakas, G., & Vassiliadis, C. A. (2017a). A report on museum branding literature. In A. Kavoura, D. Sakas, & T. P. (Eds.), *Strategic innovative marketing* (pp. 229–234). Springer. https://doi.org/10.1007/978-3-319-56288-9_31

Belenioti, Z.-C., Tsourvakas, G., & Vassiliadis, C. A. (2019a). Do social media affect museums' brand equity? An exploratory qualitative study. *Strategic Innovative Marketing and Tourism*, 533–540. https://doi.org/10.1007/978-3-030-12453-3_61

Belenioti, Z.-C., Tsourvakas, G., & Vassiliadis, C. A. (2019b). Museums brand equity and social media. *Strategic Innovative Marketing and Tourism*, 1215–1222. https://doi.org/10.1007/978-3-030-12453-3_139

Belenioti, Z. C., & Vassiliadis, C. A. (2016). Branding in the new museum era. In *Strategic innovative marketing* (pp. 115–121). Springer. https://doi.org/10.1007/978-3-319-33865-1

Belenioti, Z.-C., & Vassiliadis, C. A. (2017). Social media impact on NPO brand equity: Conceptualizing the trends and prospects EuroMed academy of business. In D. Vrontis & E. T. Yaakov Weber (Eds.), *10th annual Euromed academy of business (EMAB) conference* (pp. 150–160). EuroMed Press. http://www.emrbi.org/bop2017.pdf.

Bihamta, H., Jayashree, S., Rezaei, S., Okumus, F., & Rahimi, R. (2017). Dual pillars of hotel restaurant food quality satisfaction and brand loyalty. *British Food Journal, 119*(12), 2597–2609.

Bounfour, A. (2016). *Digital futures, digital transformation.* https://doi.org/10.1007/978-3-319-23279-9_4

Bran, E., Bautu, E., & Popovici, D. M. (2020). Towards a sustainable future: Ubiquitous knowledge mixed reality museum. *Procedia Computer Science, 176*, 2878–2885. Elsevier B.V. https://doi.org/10.1016/j.procs.2020.09.266

Bruns, A. (2012). Reconciling community and commerce? *Information, Communication & Society, 15*(6), 815–835. https://doi.org/10.1080/1369118X.2012.680482

Buhalis, D. (2020). Technology in tourism-from information communication technologies to eTourism and smart tourism towards ambient intelligence tourism: A perspective article. *Tourism Review, 75*(1), 267–272. https://doi.org/10.1108/TR-06-2019-0258

Buhalis, D. (2022). Tourism management and marketing in transformation: Preface. In D. Buhalis (Ed.), *Encyclopedia of tourism management and marketing*. Edward Elgar Publishing.

Buhalis, D., & Inversini, A. (2014). Tourism branding, identity, reputation co-creation, and word-of-mouth in the age of social media. In *Tourism management, marketing and development volume 1: The importance of networks and ICTs* (pp. 15–40). https://doi.org/10.1057/9781137354358

Buhalis, D., & Sinarta, Y. (2019). Real-time co-creation and nowness service: Lessons from tourism and hospitality. *Journal of Travel & Tourism Marketing, 36*(5), 563–582. Routledge. https://doi.org/10.1080/10548408.2019.1592059

Burke, V., Jørgensen, D., & Jørgensen, F. A. (2020). Museums at home: Digital initiatives in response to COVID-19. *Norsk museumstidsskrift, 6*(02), 117–123. https://doi.org/10.18261/issn.2464-2525-2020-02-05

Bushey, J. (2023). *A participatory archives approach to fostering connectivity, increasing empathy, and building resilience during the COVID-19 pandemic* (pp. 2379–2393).

Calvert, C. L., et al. (2020, May). *Social media during COVID-19: A quick guide for Canadian museums.*

Casais, B. (2020). *Pilgrim co-creation, 6.*

Celuch, K. (2021). Hashtag usage and user engagement on Instagram: The case of # foodfestivals, *21*(2), 966–973. https://doi.org/10.7752/jpes.2021.s2120

Cerquetti, M. (2018). Value creation in industrial heritage management. Evidence from the City of Paper (Fabriano, Italy). *Budownictwo i Architektura, 16*(4), 035–047. https://doi.org/10.24358/bud-arch_17_164_03

Cerquetti, M., & Ferrara, C. (2018). Marketing research for cultural heritage conservation and sustainability: Lessons from the field. *Sustainability (Switzerland), 10*(3). https://doi.org/10.3390/su10030774

Colbert, F., & Courchesne, A. (2012). Critical issues in the marketing of cultural goods: The decisive influence of cultural transmission. *City, Culture and Society, 3*(4), 275–280. https://doi.org/10.1016/j.ccs.2012.11.006

Corte, V. D. (2014). Open, user, and smart innovation in cultural firms. In *Handbook of research on management of cultural products: E-relationship marketing and accessibility perspectives* (pp. 200–224). https://doi.org/10.4018/978-1-4666-5007-7.ch010

Course, M., & Systems, T. (2020). *Tourism facing a pandemic: From crisis to recovery (edited by Federica Burini) | greg richards – Academia.edu, Tourism facing a pandemic: From crisis to recovery.* https://doi.org/10.6092/978-88-97235-04-4

Darmayanti, D. (2023). Patient brand value co-creation: A conceptual framework patient brand value co-creation: A conceptual framework. In *Proceedings of the 3rd Asia Pacific international conference on industrial engineering and operations management, Johor Bahru, Malaysia, September 13–15, 2022* (pp. 3389–3396). IEOM Society International.

Del Chiappa, G. (2022, October). Tourism, hospitality and culture 4.0: Shifting towards the metaverse. In *Tourism, hospitality and culture 4.0: Shifting towards the metaverse.*

Dey, B. L., Yen, D., & Samuel, L. (2020). Digital consumer culture and digital acculturation. *International Journal of Information Management, 51*(July 2019), 102057. Elsevier. https://doi.org/10.1016/j.ijinfomgt.2019.102057

Dwivedi, Y. K., Hughes, L., Wang, Y., Alalwan, A. A., Ahn, S. J., Balakrishnan, J., Barta, S., Belk, R., Buhalis, D., Dutot, V., Felix, R., Flavián, R. F. C., Gustafsson, A., Hinsch, C., & Svend, J. W. (2022). Metaverse marketing: How the metaverse will shape the future of consumer research and practice. *Psychology and Marketing, 4*(November), 750–776. https://onlinelibrary.wiley.com/doi/10.1002/mar.21767

EU. (2014). *Report on the role of public arts and cultural institutions in the promotion of cultural diversity and intercultural dialogue 2011–2014.*

Fletcher, A., & Lee, M. J. (2012). Current social media uses and evaluations in American museums. *Museum Management and Curatorship, 27*(5), 505–521. https://doi.org/10.1080/09647775.2012.738136

Fotiadis, A. (2018). *Munich personal RePEc archive modelling wedding marketing strategies: An fsQCA analysis modelling wedding marketing strategies: An fsQCA analysis.* (88249).

Fotiadis, A., Stylos, N., & Vassiliadis, C. A. (2020). Travelling to compete: Antecedents of individuals' involvement in small-scale sports events. *Tourism Recreation Research, 0*(0), 1–17. Taylor & Francis. https://doi.org/10.1080/02508281.2020.1808934

France, C., et al. (2018). Customer brand co-creation behavior: Conceptualization and empirical validation. *Marketing Intelligence and Planning, 36*(3), 334–348. https://doi.org/10.1108/MIP-10-2017-0266

Froeschl, G., Hazan, S., & Lasaracina, K. (2008, September). Handbook on cultural web user interaction first edition (September 2008). *Minerva.*

Gannon, M., Taheri, B., Thompson, J., Rahimi, R., & Okumus, B. (2022). Investigating the effects of service recovery strategies on consumer forgiveness and post-trust in the food delivery sector. *International Journal of Hospitality Management, 107*, 103341.

Gkrimpizi, T., & Peristeras, V. (2022). Digital transformation in higher education institutions, higher education. *Association for Computing Machinery.* https://doi.org/10.1201/9781003021230-2.

Gretzel, U., et al. (2020). e-tourism beyond COVID-19: A call for transformative research. *Information Technology & Tourism, 22*(2), 187–203. Springer. https://doi.org/10.1007/s40558-020-00181-3

Hausmann, A. (2012a). Creating "buzz": Opportunities and limitations of social media for arts institutions and their viral marketing. *International Journal of Nonprofit and Voluntary Sector Marketing, 17*, 173–182. https://doi.org/10.1002/nvsm

Hausmann, A. (2012b). The importance of word of mouth for museums: An analytical framework. *International Journal of Arts Management, 14*(3).

Hesmondhalgh, D., & Pratt, A. C. (2005). Cultural industries and cultural policy. *International Journal of Cultural Policy, 11*(1), 1–13. https://doi.org/10.1080/10286630500067598

Hind, K. (2022). 'Saudi vision 2030: Applying a sustainable smart techno-cultural assessment method to evaluate museums' performance post-COVID-19. *Journal of Sustainable Development, 15*(3), 46. https://doi.org/10.5539/jsd.v15n3p46

Hoksbergen, E., & Insch, A. (2016). Facebook as a platform for co-creating music festival experiences. The case of New Zealand's Rhythm and Vines New Year's Eve festival. *International Journal of Event and Festival Management, 7*(2), 84–99. https://doi.org/10.1108/02656710210415703

Hu, M., Pantano, E., & Stylos, N. (2023). "Home alone" no more: How does the internet of things (IoT) enhance travellers' subjective well-being. *Technological Forecasting and Social Change, 192*. https://doi.org/10.1016/j.techfore.2023.122563

Hudson, S., & Hudson, R. (2013). Engaging with consumers using social media: A case study of music festivals. *International Journal of Event and Festival Management, 4*(3), 206–223. https://doi.org/10.1108/IJEFM-06-2013-0012

Hudson, S., et al. (2015). The effects of social media on emotions, brand relationship quality, and word of mouth: An empirical study of music festival attendees. *Tourism Management, 47*, 68–76. Elsevier Ltd. https://doi.org/10.1016/j.tourman.2014.09.001

Hume, M., & Mills, M. (2011). Building the sustainable iMuseum: Is the virtual museum leaving our museums virtually empty? *International Journal of Nonprofit and Voluntary Sector Marketing, 16*, 275–289. https://doi.org/10.1002/nvsm.425

Kahn, R. (2020). Corona as curator: How museums are responding to the pandemic. *Elephant in the Lab.*

Kane, G. C., et al. (2015). Strategy, not technology, drives digital transformation. MITSloan. *Management Review*, 1–25.

Kim, A. J., & Ko, E. (2012). Do social media marketing activities enhance customer equity? An empirical study of luxury fashion brand. *Journal of Business Research, 65*(10), 1480–1486. Elsevier B.V. https://doi.org/10.1016/j.jbusres.2011.10.014

Koseoglu, M. A., Rahimi, R., Okumus, F., & Liu, J. (2016). Bibliometric studies in tourism. *Annals of Tourism Research, 61*, 180–198.

Kotler, N. (2001). New ways of experiencing culture: The role of museums and marketing implications. *Museum Management and Curatorship, 19*(4), 417–425. https://doi.org/10.1080/09647770100801904

Lee, D. H. (2019). Effects of key value co-creation elements in the healthcare system: Focusing on technology applications. *Service Business, 13*(2), 389–417. Springer. https://doi.org/10.1007/s11628-018-00388-9

Lemon, K. N., & Verhoef, P. C. (2016). Understanding customer experience throughout the customer journey. *Journal of Marketing, 80*(6), 69–96. https://doi.org/10.1509/jm.15.0420

Lepouras, G., & Vassilakis, C. (2004). Virtual museums for all: Employing game technology for edutainment. *Virtual Reality, 8*(2), 96–106. https://doi.org/10.1007/s10055-004-0141-1

Li, J., Gopal, R., & Sigappi, A. N. (2021). IoT in a museum for interactive experience design. *Annals of Operations Research*. https://doi.org/10.1007/s10479-021-04419-z

Martins, P., et al. (2020). Co-creation fashion brands: A case study. In *Global Fashion Conference*. http://gfc-conference.eu/wp-content/uploads/2020/11/MARTINS-ET-AL_Co-creation-fashion-brands-a-case-study.pdf

Marty, P. F. (2008). Museum websites and museum visitors: Digital museum resources and their use. *Museum Management and Curatorship, 23*(1), 81–99. https://doi.org/10.1080/09647770701865410

Marty, P. F. (2011). My lost museum: User expectations and motivations for creating personal digital collections on museum websites. *Library & Information Science Research*, 211–219. https://doi.org/10.1016/j.lisr.2010.11.003

Mergel, I., Edelmann, N., & Haug, N. (2019). Defining digital transformation: Results from expert interviews. *Government Information Quarterly, 36*(4), 101385. Elsevier. https://doi.org/10.1016/j.giq.2019.06.002

Merz, M. A., Zarantonello, L., & Grappi, S. (2018). How valuable are your customers in the brand value co-creation process? The development of a Customer Co-Creation Value (CCCV) scale. *Journal of Business Research, 82*(August 2016), 79–89. Elsevier. https://doi.org/10.1016/j.jbusres.2017.08.018

Mighali, V., et al. (2015). Innovative IoT-aware services for a smart museum. In *WWW 2015 companion – Proceedings of the 24th international conference on world wide web* (pp. 547–550). https://doi.org/10.1145/2740908.2744711

Nambisan, S., Wright, M., & Feldman, M. (2019). The digital transformation of innovation and entrepreneurship: Progress, challenges and key themes. *Research Policy, 48*(8), 103773. Elsevier. https://doi.org/10.1016/j.respol.2019.03.018

NEMO. (2020). *Survey on the impact of the COVID-19 situation on museums in Europe* (p. 21). https://www.ne-mo.org/fileadmin/Dateien/public/NEMO_documents/NEMO_COVID19_Report_12.05.2020.pdf

Nkoulou Mvondo, G. F., et al. (2022). Impact of international tourists' co-creation experience on brand trust, brand passion, and brand evangelism. *Frontiers in Psychology, 13*(March). https://doi.org/10.3389/fpsyg.2022.866362

Ohridska-olson, R. (2015). *Cultural tourism: Definitions and typology – A research note 1*. Available at SSRN 3318843.

Okoeguale, A. (2023). Brand value co-creation revisited: A critical review of studies. *Academic Journal of Interdisciplinary Studies, 12*(2), 323. https://doi.org/10.36941/ajis-2023-0052

Ou, J. (2020). China science and technology museum boosting fight against COVID-19. *Museum Management and Curatorship, 35*(3), 227–232. Taylor & Francis. https://doi.org/10.1080/09647775.2020.1762361

Palumbo, F., Dominici, G. and Basile, G. (2014) 'The culture on the palm of your hand: How to design a user oriented mobile app for museums', *Handbook of research on management of cultural products: E-relationship marketing and accessibility perspectives*, (December 2012), pp. 225–244. doi: https://doi.org/10.4018/978-1-4666-5007-7.ch011.

Pantano, E., & Stylos, N. (2020). The Cinderella moment: Exploring consumers' motivations to engage with renting as collaborative luxury consumption mode. *Psychology and Marketing, 37*(5), 740–753. https://doi.org/10.1002/mar.21345

Practice, C. (2020). *Digital engagement strategies in cultural sector during COVID-19 June 2020 design & creative practice*.

Profile, S. E. E. (2013). Editorial Advisory Board. In *Principles of cloning: Second edition*. https://doi.org/10.1016/B978-0-12-386541-0.00053-9

Quinton, S., & Fennemore, P. (2013). Missing a strategic marketing trick? The use of online social networks by UK charities. *International Journal of Nonprofit and Voluntary Sector Marketing, 51*(October 2012), 36–51. https://doi.org/10.1002/nvsm.1450

Rahimi, R. (2017). Customer relationship management (people, process and technology) and organizational culture in hotels: Which traits matter? *International Journal of Contemporary Hospitality Management, 29*(5), 1380–1402.

Rahimi, R., Akgunduz, Y., & Bilgihan, A. (2023). Impact of COVID-19 on mental health and career anxiety of hospitality and tourism students in the UK. *Journal of Hospitality and Tourism Insights, 6*(2), 892–911.

Rahimi, R., & Gunlu, E. (2016). Implementing customer relationship management (CRM) in hotel industry from organizational culture perspective: Case of a chain hotel in the UK. *International Journal of Contemporary Hospitality Management, 28*(1), 89–112.

Rahimi, R., Köseoglu, M. A., Ersoy, A. B., & Okumus, F. (2017). Customer relationship management research in tourism and hospitality: A state-of-the-art. *Tourism Review, 72*(2), 209–220.

Rahimi, R., & Kozak, M. (2017). Impact of customer relationship management on customer satisfaction: The case of a budget hotel chain. *Journal of Travel & Tourism Marketing, 34*(1), 40–51.

Raimo, N., et al. (2021). Digitalization in the cultural industry: Evidence from Italian museums. *International Journal of Entrepreneurial Behavior & Research*. https://doi.org/10.1108/ijebr-01-2021-0082

Rezaei, S., Shahijan, M. K., Valaei, N., Rahimi, R., & Ismail, W. K. W. (2018). Experienced international business traveller's behavior in Iran: A partial least squares path modelling analysis. *Tourism and Hospitality Research, 18*(2), 163–190.

Riche Resources, EU. (2014). https://resources.riches-project.eu/glossary/cultural-institutions/

San Cornelio, G., & Cruz, E. G. (2014). Co-creation and participation as a means of innovation in new media: An analysis of creativity in the photographic field. *International Journal of Communication*, 8–20.

Schallmo, D. R. A., & Williams, C. A. (2018). *History of digital transformation* (pp. 3–8). https://doi.org/10.1007/978-3-319-72844-5_2

Schweibenz, W. (1998). The "virtual museum": New perspectives for museums to present objects and information using the internet as a knowledge base and communication system. *ISI*, (November), 185–199. Accessed October 10, 2014, from http://www.informationswissenschaft.org/wp-content/uploads/isi/isi1998/14_isi-98-dv-schweibenz-saarbruecken.pdf

Schweibenz, W. (2004). The idea of becoming: The development of virtual museums, (3), 66041.

Schwertner, K. (2017). Digital transformation of business. *Trakia Journal of Science, 15*(Suppl.1), 388–393. https://doi.org/10.15547/tjs.2017.s.01.065

Sigala, M. (2020). Tourism and COVID-19: Impacts and implications for advancing and resetting industry and research. *Journal of Business Research, 117*(June), 312–321. Elsevier. https://doi.org/10.1016/j.jbusres.2020.06.015

Sigala, M., Rahimi, R., & Thelwall, M. (2019). Big data and innovation in tourism, travel, and hospitality. *Culinary Science & Hospitality Research, 23*(6), 1–11.

Simone, C., Cerquetti, M., & La Sala, A. (2021). Museums in the Infosphere: Reshaping value creation. *Museum Management and Curatorship, 36*(4), 322–341. Taylor & Francis. https://doi.org/10.1080/09647775.2021.1914140

Stylos, N. (2020). Technological evolution and tourist decision-making: A perspective article. *Tourism Review, 75*(1), 273–278.

Stylos, N., Fotiadis, A. K., Shin, D. D., & Huan, T. C. T. (2021b). Beyond smart systems adoption: Enabling diffusion and assimilation of smartness in hospitality. *International Journal of Hospitality Management, 98*. https://doi.org/10.1016/j.ijhm.2021.103042

Stylos, N., Zwiegelaar, J., & Buhalis, D. (2021a). Big data empowered agility for dynamic, volatile, and time-sensitive service industries: The case of tourism sector. *International Journal of Contemporary Hospitality Management, 33*(3), 1015–1036. https://doi.org/10.1108/IJCHM-07-2020-0644

Suárez, L. M. C., Núñez-Valdés, K., & Alpera, S. Q. Y. (2021). A systemic perspective for understanding digital transformation in higher education: Overview and subregional context in latin america as evidence. *Sustainability (Switzerland), 13*(23). https://doi.org/10.3390/su132312956

Tardivo, G., Scilla, A., & Viassone, M. (2014). A new frontier in the satisfaction of the cultural tourist: The QR code. In *Handbook of research on management of cultural products: E- relationship marketing and accessibility perspectives* (pp. 348–366). https://doi.org/10.4018/978-1-4666-5007-7.ch017

Tommasetti, A., et al. (2010). Internet of Things and value co-creation in the cultural industry towards Service Dominant logic perspective Aurelio, (1995), 1995–1996.

UNESCO. (2009). *2009 UNESCO framework for cultural statistics*. UNESCO Institute for Statistics. http://www.uis.unesco.org

UNESCO. (2014). *Culture for development indicators (CDIS). UNESCO culture for development indicators* (pp. 132–140).

Vassiliadis, C., & Belenioti, Z. C. (2017). Museums & cultural heritage via social media: An Integrated Literature Review. *Tourismos: An International Multidisciplinary Journal for Tourism, 12*(3), 97–132. Special Issue 2017. http://www.chios.aegean.gr/tourism/vol12no3.html

Vassiliadis, C. A., & Fotiadis, T. (2008). Multiple museum construct motivators: A multivariable analysis with repertory grid analysis (RGA) approach. *Journal of Tourism, 3*(25373), 12–35. http://mpra.ub.uni-muenchen.de/25373/.

Vassiliadis, C. A., & Fotiadis, A. (2016). *Managing sport tourism experiences: Blueprinting service encounters* (pp. 195–215).

Vassiliadis, C. A., Mombeuil, C., & Fotiadis, A. K. (2021). Identifying service product features associated with visitor satisfaction and revisit intention: A focus on sports events. *Journal of Destination Marketing and Management, 19*(January), 100558. Elsevier Ltd. https://doi.org/10.1016/j.jdmm.2021.100558

Vatankhah, S., Darvishmotevali, M., Rahimi, R., Jamali, S. M., & Ebrahim, N. A. (2023). Assessing the application of multi-criteria decision making techniques in hospitality and tourism research: A bibliometric study. *International Journal of Contemporary Hospitality Management, 35*(7), 2590–2623.

Verhoef, P. C., et al. (2021). Digital transformation: A multidisciplinary reflection and research agenda. *Journal of Business Research, 122*(September 2019), 889–901. https://doi.org/10.1016/j.jbusres.2019.09.022

Vial, G. (2019). Understanding digital transformation: A review and a research agenda. *Journal of Strategic Information Systems, 28*(2), 118–144. https://doi.org/10.1016/j.jsis.2019.01.003

Wang, S. M. (2018). Effects of interactive marketing on value co-creation in cultural tourism. *Journal of Interdisciplinary Mathematics, 21*(2), 489–497.

Weilenmann, A., Hillman, T., & Jungselius, B. (2013). Instagram at the Museum: Communicating the museum experience through social photo sharing. In *CHI 2013: Changing perspectives* (pp. 1843–1852). https://doi.org/10.1145/2470654.2470689

Yi, Y., & Gong, T. (2013). Customer value co-creation behavior: Scale development and validation. *Journal of Business Research, 66*(9), 1279–1284. Elsevier Inc. https://doi.org/10.1016/j.jbusres.2012.02.026

Yu, D., & Yao, W. (2023). *Research on holographic display and technology application of art museum based on immersive design Research on holographic display and technology application of art museum based on immersive design*. https://doi.org/10.1088/1742-6596/2425/1/012048

3

Digital Marketing for the Hospitality and Tourism Industry in Industry 4.0

Imran Maqbool Khan

Introduction

Industry 4.0 or the fourth industrial revolution is an era of industrialization where human works with AI-enabled machines (Acikgoz et al., 2023, 2024; Stylos, 2023). In the fourth industrial revolution, there is an interconnection between the virtual and physical world and companies have connected their products and processes to achieve higher value for their customers. Technological advancement in Industry has disrupted almost every industry including the tourism and hospitality industry.

Digital marketing is growing rapidly in the industry 4.0, and it has changed the way tourism and hospitality companies do businesses. Tourism businesses are driving booking leads mainly from websites and social media influencers (Szymkowiak et al., 2021; Gannon et al., 2022; Bihamta et al., 2017; Rezaei et al., 2018) as the tourists have a better

I. M. Khan (✉)
Birmingham City University, Birmingham, UK
e-mail: imran.khan2@bcu.ac.uk

N. Stylos et al. (eds.), *Contemporary Marketing Management for Tourism and Hospitality*, https://doi.org/10.1007/978-3-031-65049-9_3

access to digital devices to explore online channels. This trend has transformed the digital marketing approaches businesses use to deal with customers.

Digital Marketing 4.0

In Industry 4.0, digital marketing has advanced in multiple ways:

1. Marketers are using an *omnichannel approach* to provide an excellent experience to their customers with the help of the *Internet of Things (IoTs).* In omnichannel marketing, marketers provide a seamless experience to their customers across multiple platforms such as through websites, social media platforms, voice assistants, chatbots, and apps.
2. *Artificial intelligence (AI)* has transformed digital marketing massively. It is helping marketers to offer more personalized Products and services according to the customers' needs. AI helps marketers to obtain data from different online and offline sources to identify and target the key customer segment.
3. AI also helps marketers to use advanced *data analytics* to see the customer journey. This helps marketers to serve their customers efficiently across different customer journey stages.
4. AI and machine learning help marketers to use *predictive analytics* to understand customer data and future market trends in a more effective way.
5. Finally, in this digital era marketers can *automate* their marketing campaigns on different platforms such as organic and paid social media and search engine campaigns.

In Industry 4.0, tourism and hospitality marketers are taking advantage of digital marketing for customer acquisition and customer retention (Stylos et al., 2021, b; Stylos & Vassiliadis, 2023; Rahimi, 2017; Rahimi et al., 2017; Vatankhah et al., 2023). Digital marketing has become a key part of tourism and hospitality businesses to provide their customers with a seamless experience and to gain a competitive advantage. There are several digital marketing approaches businesses are using

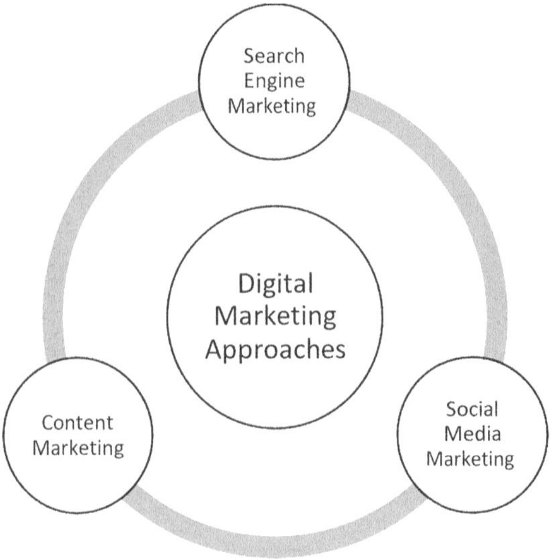

Fig. 3.1 Key digital marketing approaches for hospitality and tourism industry (Source: the Author)

to acquire new customers, retain existing customers, and provide a great customer experience. There are three key digital marketing approaches in the tourism and hospitality industry, as appearing in Fig. 3.1.

Search Marketing

Search marketing is one of the key digital marketing techniques tourism and hospitality businesses are using for customer acquisition and retention. Search marketing is the use of search engines such as Google, Bing, Yahoo, and Baidu to create awareness and bring more customers to the website of the business. Google is a leading search engine with 84.69% market share in 2023 and crucial for search marketing for businesses in the hospitality and tourism industry. It is highly competitive for businesses to appear in the above-the-fold section in Google search engine result pages (SERPs). Therefore, to appear in the above-the-fold section companies use paid and organic ways, as per Table 3.1.

Table 3.1 Search engines market shares (Source: Statista, 2023a, b)

Search Engine	Market share (%)
Google	84.69
Bing	8.85
Yahoo	2.59
Baidu	0.54

Search engines' algorithm look at different factors before presenting results to their users. For example, they can look at:

1. The *user's experience* on the website is crucial. If your website has been providing excellent user experience consistently then your visibility and ranking will increase in search engines.
2. Search engines also look at the *loading speed* of your web pages. If pages load faster, then this is a positive ranking factor.
3. *Relevance* is another crucial ranking factor. If your pages are relevant to what customers are looking for then your ranking, chances increase.

There are two ways businesses gain visibility in search engines (see Fig. 3.2):

1. Search engine optimization (SEO) (Organic)
2. Search engine marketing (Pay-per-click) (Paid)

Search Engine Optimization (SEO)

Search engine optimization (SEO), also called organic search, is one of the best digital marketing tools available in Industry 4.0. It is an organic (free) way to improve the visibility of the website of a business. Without paying anything to the search engine businesses improve the ranking of the website in search engine result pages (SERPs). There are different SEO features that can help hospitality and tourism businesses in business growth:

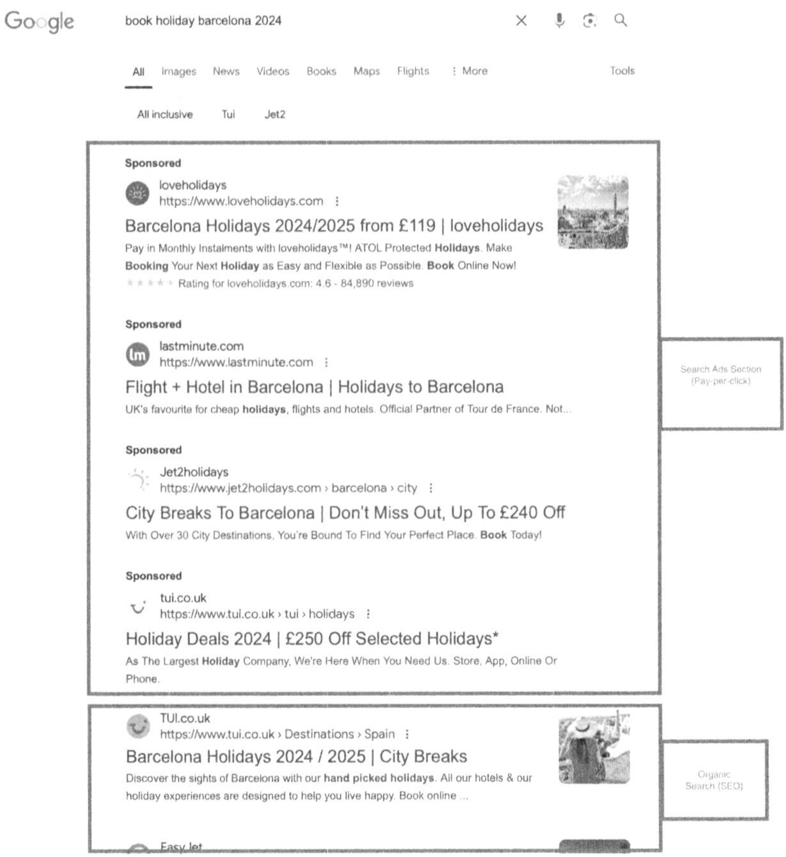

Fig. 3.2 Search ads and SEO sections in SERPs

1. *Search ranking:* It is imperative for brand awareness and visibility of the business. It helps customers to find the business while they are products and services to buy. The higher your website's ranking is in search engine result pages (SERPs) the more likely customers will land on your website to explore more about your products.
2. *Website Traffic/Sales:* Website performance depends on SEO performance. If the SEO is bad, then that will affect the website traffic as well. Customers either will not get to the website or if they do then they will not be able to complete the purchase.

3. *Trust:* A good SEO strategy will make sure that the (Readability) content on the website is relevant and easy to understand; (Usability) it is easier for customers to navigate through the website, and finally the internal and external links on the website are sound. All of this improves the website's credibility in form of website authority in search engines like Google and in the consumers' eyes as a trusted website.
4. *Cost-effective/Long-term:* SEO is cost-effective in comparison with Pay-per-Click (PPC). This is a long-term solution for the growth of any business unlike PPC; however, SEO can take some time. You should be patient when talking about the ranking of your website.

Types of SEO

There are five types of Search Engine Optimization as appearing in Fig. 3.3:

1. *On-Page SEO*

   On-page SEO is a process of optimizing the website of the business. In simple language whatever you do on your web pages to improve the ranking of your website in search engine result pages (SERPs). In this process, tourism businesses can take the following steps:

   (a) *Keyword research*: The website content is relevant to the buyer's persona. One way to make the content relevant is the use of the right keywords on website pages

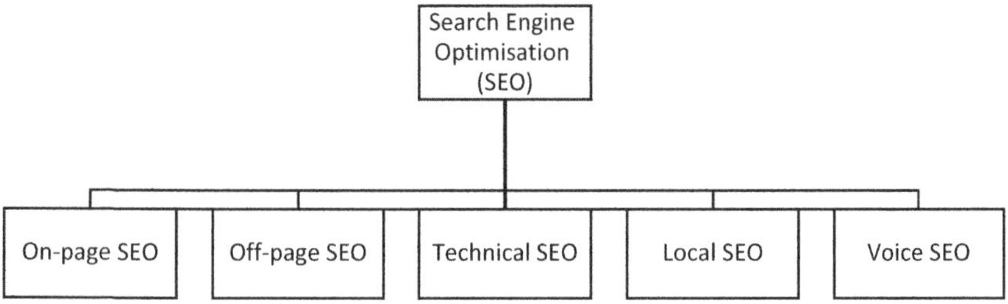

Fig. 3.3 Types of SEO (Source: the Author)

There are different types of keywords:

- *Short-tail* keywords are generic keywords or search terms consumers use when they are looking for products and services such as hotels, holidays, etc. These keywords have normally high monthly search volume and high keyword difficulty as well. That means these keywords are very competitive and it will be incredibly difficult to rank on the first page of search engines with these keywords.

There are several reasons why hospitality and tourism businesses would use short-tail keywords:

- Short-tail keywords are great for *mass marketing*. They help businesses reach a *broader audience.*
- Hospitality and tourism businesses can use these keywords to create brand awareness.
- Businesses can gain a *competitive advantage* with the use of short-tail keywords as these will target those customers who are not aware of the presence of the business.

- *Long-tail* is more focused and less competitive than short-tail keywords. These keywords are crucial particularly in the hospitality and tourism industry for several reasons:

  - Long-tail keywords can help hospitality and tourism businesses to focus on *specific target audiences.* These keywords tourism marketers use remarketing to convert their customers.
  - As long-tail keywords have *less keyword difficulty* score, therefore, chances of a hospitality business *ranking* for search results in search engine result pages (SERPs) *increase.*
  - Long-tail keywords tend to *answer customers' pain points (problems)* better than short-tail keywords. As these are more focused thus provide answers to customer questions regarding hospitality and tourism services.
  - Hospitality and tourism businesses use long-tail keywords to *enhance their targeting* such as five-star holidays in (city), pet-friendly hotels in (city), and sea-view hotels in (city).

2. *Off-Page SEO*

Unlike On-page SEO, this is a process of improving the ranking of your website in search engine result pages (SERPs) by performing different activities outside your website. In simple language, whatever you do off your website to improve your website's ranking.

There are several steps hospitality and tourism businesses can take to improve their ranking:

(a) *Link building:* Link building is one of the best strategies to do off-page SEO. This is a process of getting traffic from other websites which are dealing in the hospitality and tourism field but sell different but relevant products. For example, British Airways sells flights and holidays and any travel blogging website could be a useful link to send traffic to British Airways' website. Therefore, managing the backlink profile of a business is crucial and indeed a very challenging task. There are different ways hospitality and tourism businesses can develop backlinks or improve their existing backlink profile.

According to SEMrush, businesses can take the following steps to improve the backlink profile, as shown in Table 3.2.

(b) *Social Media Marketing:* Hospitality and tourism businesses can promote their websites on different social media websites to get more traffic to their websites. They can do both organic social media marketing and paid social media marketing to create brand awareness (Chu et al., 2020). Social media marketing really helps in developing quality backlinks. We will discuss social media marketing in more detail in the context of hospitality and tourism later in this chapter.
(c) *Content Marketing:* Content marketing is an excellent way to create brand awareness. Relevant content will make other businesses and customers share it with other potential customers that as a result helps develop a sound backlink profile.
(d) *Guest blogging:* This is an excellent way to let people know about your business. You can write blogs for other websites about hospitality and

Table 3.2 Backlink profile development (Source: SEMRush, 2023a, b)

1. Backlink quality audit	Perform a complete audit of backlinks and see if there are bad-quality links that may affect the website's ranking. Try to maintain the quality of backlinks.
2. Lost backlinks	Through your audit, you can find out any quality backlinks you lost. You can try obtaining these links again to improve the ranking of your website.
3. New backlinks	If you have recently received any new backlinks, then you can look at your recent marketing activities which helped you in gaining these links.
4. Competitors analysis	Competitor analysis can help you identify some quality links in your industry.
5. New content	New content ideas which answer the questions customers have and attract new backlinks. For example, your business can write blogs which are helpful.
6. Redirects	You can identify quality links to your previous successful web pages which you do not have any more. You redirect those links to your new web pages.

tourism and including a link back to your website helps bring the audience to your website. This will also improve the authority score of your website.

3. *Technical SEO*

Technical SEO refers to the process of fixing the technical issues of your website. If your website is technically sound, then it will help your website rank higher in search engine result pages (SERPs) and improve web traffic (Edwin, 2022). As a business you make sure that the search engine can crawl to all the web pages you have. Technical SEO is a part of On-page SEO that focuses on improving different elements on web pages to improve visibility and ranking in search engines.

There are several steps hospitality and tourism businesses can take to optimize the technical SEO of your website:

(a) *Loading Speed:* We are moving into Industry 5.0 and most people have access to 5G internet which makes shopping browsing fast whether that is on a website or social media. If customers are looking for products on your website, then it needs to load fast. In 2016 according to Marketingdive, 53% of people will leave your website if it takes more than 3 seconds to load (Marketingdive, 2016). According to Portent (2022), a website that loads in 1 second will have a 3× higher conversion rate than a website that loads in 5 seconds. Similarly, a website that loads in 1 second will have a 5 times higher conversion rate than a website that loads in 10 seconds. When pages load in 1 second the conversion rate (CR) is almost 40%. When it is 2 seconds then CR is 34% and with 3 seconds load time the CR drops to 29%. Therefore, hospitality and tourism businesses can focus on the fast load time to keep the conversion rate high.

(b) *Web crawlers/Spiders:* Search engines use web crawlers or spiders to find relevant content on your website to match with the customer search. A clear direction for web spiders about the content on your website will positively impact the user experience and ranking factor. As a business, you can also control search engine spiders from crawling to certain pages. You can use the Robot.txt file to give spiders a clear direction about crawling to your web pages.

(c) *Dead web pages:* Customer gets frustrated when a landing page of a website does not exist. Search engines do not like dead pages because they ruin the user experience. Hospitality and tourism businesses can identify dead pages on their websites and remove them. If you have identified pages that do not exist anymore, you can use redirects and send your audience to the right new web pages.

(d) *Duplicate content:* This is a very common issue with websites to have the same content on different web pages. If different URLs of your website are showing the same content, then it can confuse search engine spiders to pick the right page for the searcher. The duplicate content issue negatively affects the user experience and ranking of the website.

4. *Local SEO*

Local SEO is a process of optimizing the website and online presence to improve the ranking for local searches. Local SEO has become imperative for the hospitality and tourism industry. For example, travellers look for hotels near the place they are visiting. Similarly, travellers look for food near the place they are visiting or staying. Therefore, the "Near me" keyword is crucial for local SEO.

There are several important factors for local SEO:

(a) If the content of Hospitality and tourism businesses is *relevant* to their customers, then chances are high that businesses will rank in local search engine result pages (SERPs). For example, how relevant is your holiday website, hotel, or restaurant to the searcher?
(b) The *distance* of your business also affects the local SEO ranking. For example, how far is your hotel or restaurant from the searcher?
(c) *Popularity* is another element that will determine your local SEO rating. For example, how famous is your business locally, the number of reviews, and word-of-mouth (WOM) will affect your local SEO a lot.

There are several strategies to implement Local SEO:

- Set up a *Google Business Profile (GBP)* to make sure that local customers can see your business when they are looking for local service providers. It is a free Google service for businesses to provide their business information such as name, web address (URL), hours, address, reviews, contact number, and directions.
- Make sure you provide complete contact information to make the customer journey easier. To improve the effectiveness of your GBP you can at least provide *UNAP* (URL, Name, Address, Phone). If your contact information is missing, then chances are high that your potential customers will move to your competitors.
- *Keyword research* is crucial for Local SEO. You should hunt and add local SEO keywords for your hospitality and Tourism businesses
- *Reviews* play an important role in your local SEO. You should manage and respond to your reviews in an effective manner.

5. *Voice SEO*

Voice SEO is a process of optimizing the website and online presence to improve the ranking for voice searches. According to Li et al. (2009), "Voice search is the technology that enables users to access information using spoken queries". The total number of voice searchers is increasing. Seventy percent of users give voice commands to search for something.

People mainly use voice search options to obtain information, for navigation and for local searches.

Voice assistants (VAs) are revolutionizing the whole hospitality and tourism experience (Rozumowski et al., 2020). Voice assistant technology uses voice-based algorithms to suggest hotels and places to visit. According to Statista (2023a, b), there will be 8.4 billion digital voice assistants by 2024 that is higher than the world's population (Statista, 2022). Hotels are now using voice assistants as support tools for their guests to improve the overall quality of the interaction and experience (Amatulli et al., 2022).

Hospitality and tourism businesses can take the following actions to make sure that their websites are ranking for voice searches. According to Semrush, there are number of techniques:

(a) Customers are always looking for quick solutions. Therefore, businesses that are answering questions as *featured snippets* will rank for voice searches. The loading speed of featured snippets is relatively quicker which is another reason why it will appear for voice searches (Rahimi & Kozak, 2017; Rahimi & Gunlu, 2016; Koseoglu et al., 2016).
(b) *Seventy-one percent* of customers now prefer voice search over text search (Pekar et al., 2024). Voice search has skyrocketed by 146%, from 3.25 billion users to 8 billion users in 2023.
(c) If *Local SEO* is right, then chances are high that websites will rank for voice search. Hospitality Businesses can add Google Business Profiles including, Addresses, Phone numbers, URLs, and Opening timings. "Near me" searches are very common for voice searches.
(d) *Reviews* can have a significant impact on voice search results. For example, when users ask voice assistants questions like, "Which hotels in London are good for weddings?" then Google highlights reviews that mention "weddings".
(e) *Site Authority Score and Quality of backlinks* can also impact on voice ranking. Search engine algorithms most probably show website for voice searches that have high authority score and high-quality backlinks.

Search Paid Advertisement

Many people search daily to find tourism- and hospitality-related products and services using multiple search engines in the following section.

Google Ads

Google is a leading search engine with 84.69% market share in 2023 and crucial for search marketing for businesses in the hospitality and tourism industry. Google Ads is a service by Google to allow businesses to place ads in Google search results and its advertising network. The sponsored search engine results are paid results in the Google search engine. Google search Ads is an online advertising program. Tourism and hospitality businesses use Google search Ads to reach potential customers with relevant products and services.

In order to create an effective and focused Google ad for your hospitality and tourism business you need to create a Google ad campaign strategy. According to Google, you can take the following steps:

1. First step for "Google ads" using any search engine. You will get to the home page. You can click on "**start now**" to start your business ad campaign.
2. *Goals:*

Setting the right *goals* is the next important step in your Google ad campaign strategy (Google Ads, 2023). You can have multiple objectives as a tourism and hospitality business:

(a) Create your brand awareness and reach.
(b) Show people your products and services.
(c) Attract the right people to visit your website.
(d) Increase your online, in-app, by-phone, or in-store sales.
(e) Get quality leads by encouraging customers to take action.
(f) Promote your app.
(g) Increase visits to your local shop.

3. Campaign Types
   Once your Google ad campaign objectives are sorted then you select a *campaign type*. You have multiple campaign types, as shown in Table 3.3.

4. *Targeting and Audience Segments*

Hospitality and tourism segmentation is a process of dividing tourists into different groups according to shared characteristics, for example, geographic, demographic, psychographic, or behavioral characteristics

Table 3.3 Types of Google ad campaigns for hospitality and tourism business (Source: Google Ads, 2023)

Search campaign	• This is a text ad on Google search results and helps you reach the high-intent customer on a Google search looking for products and services. • You can also *Google dynamic search ads* which will use the information on your website to reach potential customers who are searching on Google for products and services you offer. • If you have missed any important keywords in your keywords ads campaigns, then this can help you to fill in those gaps. • You can create block pages by adding negative keywords to your dynamic search campaigns to avoid certain audiences.
Performance max campaign	• This goal-based campaign uses automation to access all of Google ads inventory from a single campaign and builds top-performing ads. • This will use your objectives to reach potential customers by automatically showing your ads. • It will find a better target audience by complementing all your other Google Ads campaigns across all Google's channels such as YouTube, Search, Display, Discover, Gmail, and Maps (Google for Retail, 2023).
Display campaign	• This campaign helps you show your ads across 35 million sites and apps, and Google-owned properties such as YouTube and Gmail in the Google display network. • Display campaign uses machine learning for smart bidding and to optimize targeting. • Display campaigns use responsive display ads. Your business will upload a variety of assets in form of images, headlines, logos, videos, and descriptions of your products in Google Ads and the responsive ad will automatically adjust the size, appearance and format of your assets and adjust in any available ad space.

(*continued*)

Table 3.3 (continued)

Discovery campaign	• You can use this campaign type to reach 3 billion potential customers using Google's feed. • If people are ready to discover and engage with your brand, then a discovery campaign is ideal for your business. • You can show highly visually rich and personalized ads to people who ready to engage with your brand across Google properties. • Google determines the high-intent customers with the help of the Google algorithm and shows discovery ads in Google's feeds on the YouTube app, The Google App, and the Gmail app (Martina, 2022).
Shopping campaign	• This campaign shows your product to shoppers in a more visual format when they are looking for what to buy. • Shopping ads uses product attributes from Merchant Centre rather than keywords to decide where and how to show your shopping ad. • Shopping ads include the following details of your product: photo, title, price, shop name, rating, reviews, payment options, return policy, delivery options, and more. • Customers can see your products before visiting your website and get a good sense of the product in terms of design, price, and reviews. • You can also use *local inventory feature* to target local customer searching on Google.
Video campaigns	• Video campaigns allow you create and run video ads YouTube, and high-quality publisher websites and apps in Google video partners. • You can control the targeting of your video ad by using audience targeting method. In this method you define your audiences who you want to show your video ad to such as demographic groups, detailed demographics, interests, affinity segments, data segments, customer match, and similar segments options. • You can also use content targeting methods to define where you want your video ads to show such as placements (YouTube channels, YouTube videos, and websites and apps on Google display network), topics, and keywords.
Smart campaigns	• Smart campaigns are easy solutions for small businesses. • You do not need to select a bid strategy. Smart campaign will automatically manage biding for you. • It automatically creates keywords. • It will show you only the most important statistics in your smart ad campaign. • Smart campaign automatically shows your ad on Google Ads network. • Unlike other above-mentioned ads campaigns, smart campaign only allows you to link one landing page per campaign.

(Stylos, 2022). This is possible to do while you are creating your campaigns in Google Ads:

(a) This allows you to select the exact *location* where you want to show your ad to attract your potential customers. You can pick a country, city, region, or postcode as a targeted ad location. In addition to it, you can also decide the radius to further improve your reach.
(b) You can select the relevant audience segment using the audience segment. This section also allows you to create a new segment using the audience manager.
(c) Finally, you select the language your potential customers speak.

5. *Budget:* The next step is to pick your average daily budget. Businesses should consider this important decision regarding how much to spend towards search ads. Interestingly, the combined spending of online travel giants Booking Holdings and Expedia Group spent more than $4.65 billion on performance advertising and out of this $2.9 billion was on Google performance ads (Statista, 2023a, b). Booking owns Europe's www.Booking.com and U.S. businesses including Priceline and Kayak. Overall, the Travel industry spent at least $8.9 billion on Google performance advertising in 2021 and revenue increased from 21.7% to $46.2 in the fourth quarter of 2021 (Skift, 2022). Similarly, according to the Tourism Marketing Agency, Rosehill Travel has a total ad budget of £1,798 which resulted in £62, 196 to £828,712 in potential booking enquiries. This was "primarily related to the use of online search engines (Primarily Google)". This trend has increased after COVID-19 (Jiang & Stylos, 2021). For example, according to Statista, the marketing and selling expenses of Expedia Group, Inc. increased again after a sharp drop in COVID-19 to $6.1 billion in 2022 from $4.22 billion in 2021. This shows that search ads, especially good ads, are crucial in the post-pandemic world.

This is where we also need to understand a few fundamental concepts we discussed above:

- *Google Performance Ads:* This goal-based campaign uses automation to access all of Google Ads inventory from a single campaign and builds top-performing ads. This will use your objectives to reach potential customers by automatically showing your ads. It will find a better target audience by complementing all your other Google Ads campaigns across all Google's channels such as YouTube, Search, Display, Discover, Gmail, and Maps.

- *Performance Advertising:* Performance advertising is a type of digital advertising where businesses pay for measurable results.

Content Marketing

Content marketing is the process of creating relevant content and sharing it on multiple platforms with your target audience. According to Chen et al., there are three stages of consumers' online holiday decision-making process: (1) Information search, (2) Evaluation, (3) Purchase (2015).

1. *Information Search:* Social media shapes the tourists' desires and perceptions of holiday destinations. Tourists' travel photos of a specific destination on social media play an important role in destination selection and in the traffic to the destination. User-generated travel content (UGC) on social media is more influential than the advertisement campaigns by the hotels.

In this stage, customers have some idea about the product and service and Hospitality and tourism businesses can create informative content. For example, running a TV ad with the message "Holiday Season" can help people recognize the need for a holiday. Businesses normally use the push marketing approach to reach people and the focus is broad, so most people can see the content.

Hospitality and tourism businesses can make sure that their content is accessible to potential consumers. According to Think with Google, 51% of global shoppers use search to find more information about the products they are planning to buy online. At this stage, companies can make sure their website is ranking high in search engines and providing relevant and detailed informative content.

2. *Evaluation:* At this stage, the customer is engaged with the content and your product- related content is answering the questions consumers have. To convert your potential customer into your real customer you need to provide all possible information your customer might be looking at to finalize the purchase. For example, this could be about the return policy, finance options, or any product spec or tech-related queries. At this stage, potential customers might be visiting your websites and social media pages to finalize the purchase. Therefore, as a hospitality and tourism business, you need to keep your website and social media up-to-date, information should be consistent, and web pages should be user-friendly and mobile-friendly.
3. *Purchase:* At this stage as a business, you are successful in convincing your potential customers to buy your product; however, they are still in the trial phase. At this stage, you need to invest in kind of content that can retain your customers. Due to the intangible nature of the tourism product, the evaluation will be done once the service has been consumed.

Content Optimization Strategies

Content marketing is crucial for the hospitality and tourism industry. Visual content has become essential for customer experience sharing. People look for visual content published by other people before making a purchase decision. There are several factors that make tourism and hospitality content attractive and relevant:

4. *Colors: Colors* have a strong relationship with human emotion, body, and mind. Some colors provide happiness and others give a sense of urgency. Therefore, tourism and hospitality businesses should be very careful when using colors in decoration. Color also plays an important role in consumer decision-making. Colors can create brand awareness and recognition by 80% and similarly, 93% of consumers' purchase decisions are solely based on visuals. Color can shape consumers' purchase intentions, product meaning, brand personality, and product quality and price.

According to Gorn et al. (1997), there are three dimensions to coordinate color number: *Lightness, Chroma, and Hue (LCH).* According to Valdez and Mehrabian (1994), the brighter the color is, the "whiter" it appears to humans. A color that is "whitish" can have a calming and relaxing impact. Colors can have a significant impact on consumers' online shopping behavior (Jiang & Stylos, 2021). Less bright website backgrounds can lead to higher patronage intentions whereas brighter backgrounds can increase consumers' price sensitivity (Yu et al., 2020).

Tourists often make decisions based on the *user-generated content (UGC).* UGC refers to the content created by unpaid online contributors. Facebook, Instagram, and TripAdvisor are popular UGC platforms among tourists. UGC plays an important role in all stages of holiday decision-making as it is considered more reliable and neutral than information provided by official means.

5. *Marketing Appeals:* Marketing appeals play an important role in both organic and paid marketing campaigns. The right marketing appeal will possibly improve the "engagement rate" (number of people engaging with the online content) and "conversion rate" (number of people completing actions created by the business on the website or social media pages). Marketing appeals used in the content can increase or decrease the level of attention by the users. There are several marketing appeals for hospitality and tourism businesses to improve the reach, engagement, and conversion of the content:

    (a) *Rationale Appeal* is also called informational advertising appeal. These appeals are designed to attract the target audience through a rational thought process (Casais & Pereira, 2021). For example, tourism and holiday companies can highlight different features of their product package for a rational thought process such as all-inclusive features, Images, price, and related questions that can provide further information for decision-making. Adding rationale appeal in the content helps customers significantly in all stages of decision-making; however, it plays a vital role in the final stages of decision-making.

(b) *Sex appeal* in marketing is the use of erotic images or sexual words to attract customers. It is a highly controversial advertising method. According to Hu et al. (2020), people prefer high-sex appeal ads at the initial gazing stage and low-sex appeal ads at the final evaluation stages of the buying process. Sexually appealing ads are good for attracting customers at first; however, excessive sexual content can lose the attention of customers. Thus, excessive sexual content can have a negative impact on customers. The use of sex appeal is common in the hospitality and tourism industry with many hotels using photos showing women in bikinis to promote their beachfront. Another research by Cummins et al. (2021) found that men spend 40% more time looking at the model in the ad with sexual appeals. Therefore, when and where to add sexual appeal to content is crucial for businesses in the hospitality and tourism industry.

(c) *Humor Appeal*, according to Piirman et al. (2019), makes tourism and hospitality experiences fun and enjoyable. Tourism and hospitality businesses can attract new customers and gain a competitive advantage through co-creating tourism experiences and humor appeal can enhance this co-creation. Advertisers understand that feelings and emotions are conveyed through humor. Humor appeal, if applied well, can attract and engage customers and connect them to the brand. For example, in the promotion of the Spain brand, humor is expressed through skin marks (Porres-Guerrero & Foronda-Robles, 2019).

(d) *Sustainability Appeal* is a vital role in communication strategies and tourism managers understand that. If sustainability is part of the marketing campaign it can have a positive impact on consumers and the destination's competitive advantage. As the sustainability and green trends are increasing. According to a booking.com survey in 2023, 76% of travellers tend to travel more sustainably in the coming year. Recent research by Amatulli et al. (2021) found that marketing communication strategies with sustainability-focused messages will lead to a higher willingness to book a room because it enhances customers' perception of the luxury hotel's integrity.

Similarly, booking.com started a sustainability program in 2021 to promote sustainable travelling. There are sustainability levels for customer information on the booking site. Globally, over two million accommodation partners can see their sustainability levels and customers consider these levels before booking (Booking.com, 2023).

Social Media Marketing

Social media plays a vital role in hospitality and tourism businesses. For example, Instagram is considered a powerful photo-sharing platform for tourist destinations. There are 500 million daily active users, and 100 million photos are shared every day. Rapid development in social media platforms has also changed the way hospitality and tourism companies advertise and market on social media. Sharing photographs before, during, and after travel has become a trend and a new form of communication on social media (Sigala et al., 2019; Rahimi et al., 2023).

Social media marketing is the process of creating content for social media websites to create awareness about the products and services (Li et al., 2021), build a community, and drive traffic to your business (Chu et al., 2020). Social media marketing is crucial for hospitality and tourism businesses. According to Revenue Hub, 97% of millennials share photos on social media during travels and 71% of travel agencies report better interactions with consumers on social media (Revenue Hub, 2022). According to Greenlaw (2022), social media can help hospitality and tourism businesses in different ways:

- *Strategic Goals:* Businesses should have a very clear idea of how and which social media platforms will be used to achieve their strategic goals.
- *Expand Reach:* Organizations use social media marketing to expand their geographical reach to buyers.
- *Relationship Marketing:* Organizations use social media to develop and strengthen their relationship with customers.

Khan's Model of Empowerment in Hospitality and Tourism Decision-Making

Many consumers turn to social media to acquire information about products and services. Due to the development of social media, the way tourist information is accessed and used has changed dramatically. Social media empowers socially disempowered people in their holiday decision-making. Khan et al. (2022) developed an empowerment model that measures the role of social media in empowerment in holiday decision-making. Khan has found that social media communication can be associated with empowerment in hospitality and tourism decision-making; see Fig. 3.4.

- *Autonomous self-expression* is a process of expressing oneself freely on social media based on one's interests. Self-expression is not limited to sharing information on social media but also includes activities such as "likes" and sharing posts.

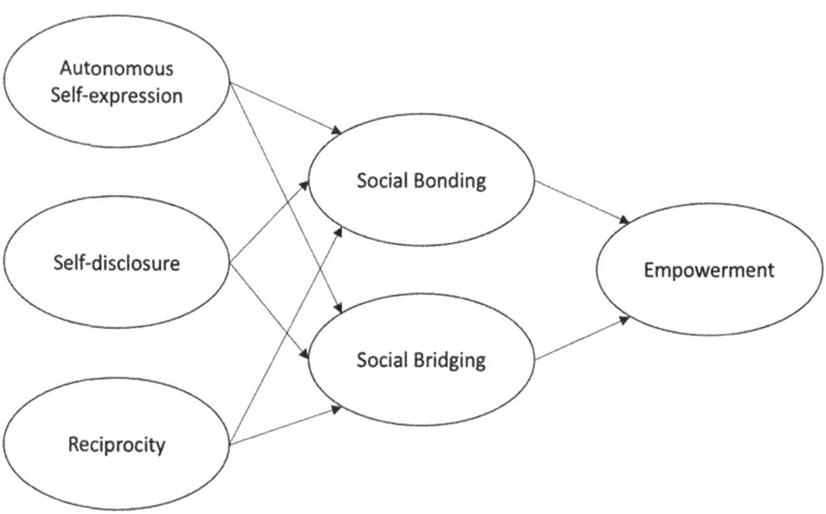

Fig. 3.4 Khan's model of empowerment (Source: Author)

- *Self-disclosure* of hospitality and tourism consumption is defined as the sharing of information about hospitality and tourism experiences, feelings, and opinions through social media. Individuals may disclose their general identity on social media and use focused self-disclosure to talk about specific topics like hospitality and tourism.
- Focused self-disclosure in tourism consumption is expected to create and improve social bonding and bridging.
- *Reciprocity* is a mutual effect and almost a concomitant outcome of self-disclosure, which means people mimic the behavior of other people on social media whom they interact with.
- *Social capital and empowerment in holiday decision-making:* Social capital is an empowering mechanism for disempowered communities. Khan et al. (2022) found that based on social power theory, the empowerment of socially disempowered people in hospitality and tourism decision-making is linked to:

  – Their ability to choose hospitality and tourism alternatives independently.
  – Their ability to influence such choices that will impact their tourism and hospitality consumption experiences in a group (e.g. family).

Khan's model of empowerment focuses on social media-induced empowerment and the role of social media usage in social capital accumulation, which in turn helps disempowered segments in holiday planning. It also confirms the power of both weak and strong ties in generating empowerment for decisions such as where to travel and where to stay.

Recommended Further Reading

Sigala, M., Rahimi, R., & Thelwall, M. (2019). Big data and innovation in tourism, travel, and hospitality. *Culinary Science & Hospitality Research, 23*(6), 1–11.

Stylos, N. (2020). Technological evolution and tourist decision-making: A perspective article. *Tourism Review, 75*(1), 273–278.

Stylos, N., Rahimi, R., Okumus, B., & Williams, S. (2021). *Generation Z marketing and management in tourism and hospitality* (pp. 06–07). Springer International Publishing.

Stylos, N., & Zwiegelaar, J. (2019). *Big data as a game changer: How does it shape business intelligence within a tourism and hospitality industry context?* (pp. 163–181). Springer.

References

Acikgoz, F., Perez-Vega, R., Okumus, F., & Stylos, N. (2023). Consumer engagement with AI-powered voice assistants: A behavioral reasoning perspective. *Psychology & Marketing, 40*(11), 2226–2243.

Acikgoz, F., Stylos, N., & Lythreatis, S. (2024). Identifying capabilities and constraints in utilizing blockchain technology in hospitality and tourism. *International Journal of Contemporary Hospitality Management.* https://doi.org/10.1108/IJCHM-07-2023-1083

Amatulli, C., De Angelis, M., & Stoppani, A. (2021). The appeal of sustainability in luxury hospitality: An investigation on the role of perceived integrity. *Tourism Management, 83*, 104228.

Amatulli, C., Sestino, A., Peluso, A. M., & Guido, G. (2022). Luxury hospitality and the adoption of voice assistants: The role of openness to change and status consumption. In *The emerald handbook of luxury management for hospitality and tourism* (pp. 285–303). Emerald Publishing Limited.

Bihamta, H., Jayashree, S., Rezaei, S., Okumus, F., & Rahimi, R. (2017). Dual pillars of hotel restaurant food quality satisfaction and brand loyalty. *British Food Journal, 119*(12), 2597–2609.

Booking.com. (2023). *Sustainability programme.* https://www.sustainability.booking.com/booking-travel-sustainable

Casais, B., & Pereira, A. C. (2021). The prevalence of emotional and rational tone in social advertising appeals. *RAUSP Management Journal, 56*, 282–294.

Chen, C. H., Nguyen, B., Klaus, P. P., & Wu, M. S. (2015). Exploring electronic word-of-mouth (eWOM) in the consumer purchase decision-making process: The case of online holidays– evidence from United Kingdom (UK) consumers. *Journal of Travel & Tourism Marketing, 32*(8), 953–970.

Chu, S. C., Deng, T., & Cheng, H. (2020). The role of social media advertising in hospitality, tourism and travel: A literature review and research agenda. *International Journal of Contemporary Hospitality Management, 32*(11), 3419–3438.

Cummins, R. G., Gong, Z. H., & Reichert, T. (2021). The impact of visual sexual appeals on attention allocation within advertisements: An eye-tracking study. *International Journal of Advertising, 40*(5), 708–732.
Edwin, T. (2022). *What's technical SEO? 8 technical aspects everyone should know*. https://yoast.com/what-is-technical-seo/
Gannon, M., Taheri, B., Thompson, J., Rahimi, R., & Okumus, B. (2022). Investigating the effects of service recovery strategies on consumer forgiveness and post-trust in the food delivery sector. *International Journal of Hospitality Management, 107*, 103341.
Google Ads. (2023). *Google ads campaigns*. https://ads.google.com/aw/campaigns/new?ocid=1222375862&cmpnInfo=%7B%228%22%3A%2256f46df2-1bf3-4728-a52e-d17db492dd2e%22%7D&subid=uk-en-et-g-aw-c-home-awhp_xin1_top%21o2-awhp-hv-01-22-ahpm-0000000126-0000000000&euid=95980997&u=9005951453&uscid=1222375862&c=9879016838&authuser=0&sourceid=emp
Google for Retail. (2023). *Performance max campaigns*. https://www.google.com/intl/en_uk/retail/solutions/performance-max/?utm_source=google&utm_medium=ha-sem&utm_campaign=bkws-1p-campaign&mcsubid=gb-en-ha-mc-bk-c-she!o3&gclid=Cj0KCQjwiZqhBhCJARIsACHHEH8Ne8KJZUWskCpXMYFO99CbrrLKRf6B132UXFtDVuYeb75YEwIKrAIaAneBEALw_wcB
Gorn, G. J., Chattopadhyay, A., Yi, T., & Dahl, D. W. (1997). Effects of color as an executional cue in advertising: They're in the shade. *Management Science, 43*(10), 1387–1400.
Greenlaw, M. (2022). *The of social media in the hotel and hospitality industry*. https://uphotel.agency/the-power-of-social-media-in-the-hotel-and-hospitality-industry/?utm_source=revenue-hub&utm_edium=referral&utm_campaign=uphotel-article-link-rh-post
Hu, F., Wu, Q., Li, Y., Xu, W., Zhao, L., & Sun, Q. (2020). Love at first glance but not after deep consideration: The impact of sexually appealing advertising on product preferences. *Frontiers in Neuroscience, 14*, 465.
Jiang, Y., & Stylos, N. (2021). Triggers of consumers' enhanced digital engagement and the role of digital technologies in transforming the retail ecosystem during COVID-19 pandemic. *Technological Forecasting and Social Change, 172*. https://doi.org/10.1016/j.techfore.2021.121029
Khan, I. M., Sahadev, S., Rashid, T., & Banerjee, S. (2022). Social media and empowerment in hospitality and tourism decision-making: A study among UK Muslim women. *International Journal of Hospitality Management, 101*, 103125.

Koseoglu, M. A., Rahimi, R., Okumus, F., & Liu, J. (2016). Bibliometric studies in tourism. *Annals of Tourism Research, 61*, 180–198.

Li, T., Bao, C., Xu, W., Pan, J., & Yan, Y. (2009). Improving voice search using forward-backward LVCSR system combination. In H. Wang, Y. Shen, T. Huang, & Z. Zeng (Eds.), *The sixth international symposium on neural networks (ISNN 2009)* (pp. 769–777). Springer.

Li, F., Larimo, J., & Leonidou, L. C. (2021). Social media marketing strategy: Definition, conceptualization, taxonomy, validation, and future agenda. *Journal of the Academy of Marketing Science, 49*, 51–70.

Marketingdive. (2016). *Google: 53% of mobile users abandon sites that take over 3 seconds to lead.* https://www.marketingdive.com/news/google-53-of-mobile-users-abandon-sites-that-take-over-3-seconds-to-load/426070/#:~:text=New%20research%20by%20Google%20has,seconds%20on%20a%204G%20connection

Martina, B. (2022). *What are Google discovery ads? Examples + a campaign tutorial.* https://blog.hubspot.com/marketing/google-discovery-ads

Pekar, V., Candi, M., Beltagui, A., Stylos, N., & Liu, W. (2024). Explainable text-based features in predictive models of crowdfunding campaigns. *Annals of Operations Research*, 1–31.

Piirman, M., Pratt, S., Smith, M., & Tooman, H. (2019). Assessing humour use in accommodation establishments: The Estonian context. *European Journal of Humour Research, 7*(3), 63–83.

Porres-Guerrero, M., & Foronda-Robles, C. (2019). Where is the humour in tourism promotion? An investigation of the "Spain Marks" campaign. *The European Journal of Humour Research, 7*(3).

Portent. (2022). *Site speed is (still) impacting your conversion rate.* https://www.portent.com/blog/analytics/research-site-speed-hurting-everyones-revenue.htm

Rahimi, R. (2017). Customer relationship management (people, process and technology) and organizational culture in hotels: Which traits matter? *International Journal of Contemporary Hospitality Management, 29*(5), 1380–1402.

Rahimi, R., Akgunduz, Y., & Bilgihan, A. (2023). Impact of COVID-19 on mental health and career anxiety of hospitality and tourism students in the UK. *Journal of Hospitality and Tourism Insights, 6*(2), 892–911.

Rahimi, R., & Gunlu, E. (2016). Implementing customer relationship management (CRM) in hotel industry from organizational culture perspective: Case of a chain hotel in the UK. *International Journal of Contemporary Hospitality Management, 28*(1), 89–112.

Rahimi, R., Köseoglu, M. A., Ersoy, A. B., & Okumus, F. (2017). Customer relationship management research in tourism and hospitality: A state-of-the-art. *Tourism Review, 72*(2), 209–220.
Rahimi, R., & Kozak, M. (2017). Impact of customer relationship management on customer satisfaction: The case of a budget hotel chain. *Journal of Travel & Tourism Marketing, 34*(1), 40–51.
Revenue Hub. (2022). *The of social media in the Hotel and Hospitality Industry.* https://revenue-hub.com/the-power-of-social-media-in-the-hotel-and-hospitality-industry/
Rezaei, S., Shahijan, M. K., Valaei, N., Rahimi, R., & Ismail, W. K. W. (2018). Experienced international business traveller's behavior in Iran: A partial least squares path modelling analysis. *Tourism and Hospitality Research, 18*(2), 163–190.
Rozumowski, A., Schäfer, W., & Klaas, M. (2020). Resistance to customer-driven business model innovations: An explorative customer experience study on voice assistant services of a Swiss tourism destination. *Athens Journal of Tourism, 7*(4), 191–208.
SEMRush. (2023a). *Voice search SEO: Top tips on how to rank for voice search.* https://www.semrush.com/local-seo/voice-search-seo/
SEMRush. (2023b). *Voice search marketing.* https://www.semrush.com/local-seo/voice-search-in-marketing/
Sigala, M., Rahimi, R., & Thelwall, M. (2019). Big data and innovation in tourism, travel, and hospitality. *Culinary Science & Hospitality Research, 23*(6), 1–11.
Skift. (2022). *Can online travel's titans ever break free of Google.* https://skift.com/2022/05/05/can-online-travels-titans-ever-break-free-of-google/
Statista. (2022). *Number of digital voice assistants in use worldwide 2019–2024.* Number of voice assistants in use worldwide 2019–2024 | Statista
Statista. (2023a). *Combined expenditure on Google performance advertising of Booking Holdings and Expedia Group worldwide from 2019 to 2021.* https://www.statista.com/statistics/1364477/expenditure-google-ads-booking-expedia-worldwide/#:~:text=Despite%20the%20sharp%20annual%20increase,travel%20giants%20on%20performance%20advertising
Statista. (2023b). *Market share of leading desktop search engines worldwide from January 2015 to March 2023.* https://www.statista.com/statistics/216573/worldwide-market-share-of-search-engines/
Stylos, N. (2022). An integrated duality theory framework (IDTF): Marking pathways for consumer decision-making researchers in the hospitality and tourism industry. *International Journal of Contemporary Hospitality Management, 34*(7), 2597–2619.

Stylos, N. (2023). Intelligence augmentation via human-AI symbiosis: Formulating wise systems for a metasociety. In *Philosophy of artificial intelligence and its place in society* (pp. 301–314). IGI Global.

Stylos, N., Fotiadis, A. K., Shin, D., & Huan, T. C. (2021). Beyond smart systems adoption: Enabling diffusion and assimilation of smartness in hospitality. *International Journal of Hospitality Management, 98*(2021), 103042. https://doi.org/10.1016/j.ijhm.2021.103042

Stylos, N., & Vassiliadis, C. A. (2023). Gamification design: Toward developing image perception scales for generation Z consumers. *International Journal of Contemporary Hospitality Management.* https://doi.org/10.1108/IJCHM-12-2022-1616

Stylos, N., Zwiegelaar, J., & Buhalis, D. (2021). Big data empowered agility for dynamic, volatile, and time-sensitive service industries: the case of tourism sector. *International Journal of Contemporary Hospitality Management, 33*(3), 1015–1036.

Szymkowiak, A., Gaczek, P., & Padma, P. (2021). Impulse buying in hospitality: The role of content posted by social media influencers. *Journal of Vacation Marketing, 27*(4), 385–399.

Valdez, P., & Mehrabian, A. (1994). Effects of color on emotions. *Journal of Experimental Psychology: General, 123*(4), 394.

Vatankhah, S., Darvishmotevali, M., Rahimi, R., Jamali, S. M., & Ebrahim, N. A. (2023). Assessing the application of multi-criteria decision making techniques in hospitality and tourism research: A bibliometric study. *International Journal of Contemporary Hospitality Management, 35*(7), 2590–2623.

Yu, C. E., Xie, S. Y., & Wen, J. (2020). Coloring the destination: The role of color psychology on Instagram. *Tourism Management, 80*, 104110.

4

Smart Technologies Use by SMEs in the Hospitality Industry: Implications for Customer Interactions and Organizational Resilience

Albena Pergelova, Shelley Beck, Jeremy Zwiegelaar, and Nikolaos Stylos

Introduction

Smart technologies (such as big data, AI, robotics, smart platforms/apps) have become important in the hospitality industry, as they allow companies to interact with a variety of stakeholders and provide better experiences to consumers, while improving efficiency (Stylos et al., 2021b, b).

A. Pergelova (✉)
MacEwan University, Edmonton, AB, Canada
e-mail: pergelovaa@macewan.ca

S. Beck • J. Zwiegelaar
Oxford Brookes University, Oxford, UK
e-mail: sbeck@brookes.ac.uk; jzwiegelaar@brookes.ac.uk

N. Stylos
University of Bristol Business School, Bristol, UK
e-mail: n.stylos@bristol.ac.uk

N. Stylos et al. (eds.), *Contemporary Marketing Management for Tourism and Hospitality*, https://doi.org/10.1007/978-3-031-65049-9_4

Industry 4.0 applications (e.g., big data, robotics, 3D printing, blockchain technologies and smart platforms/apps) have given a boost to open innovation, and call for new business models (BMs) to turn technological capabilities to new business opportunities for organizations across the spectrum of the service sector (Jiang & Stylos, 2021). SMEs in the hospitality sector have been distressed during the COVID-19 pandemic, with much uncertainty about their ability to survive. Organizational resilience has become a key concern for SMEs in the hospitality industry (Ntounis et al., 2022; El-Said et al., 2023; Rahimi & Kozak, 2017; Rahimi & Gunlu, 2016). Smart technologies have been considered to provide the potential to help SMEs for recovering from the severe damage caused globally due to the lockdowns, and to build resilience and adaptability (Stylos et al., 2021, b; Prayag et al., 2023).

Although smart technologies bring the potential for improved performance, the realization of this potential depends on the ability of SMEs to deploy technological solutions effectively. Among the major barriers for smart technology implementation by SMEs are cognitive barriers, such as reluctance to change, worries over the costs/ROI from smart technology, and environmental barriers (e.g., privacy and security legislation, consumer concerns about data use) (Chan et al., 2018; Law et al., 2021). SMEs in the hospitality industry also need to consider carefully how to best use smart technologies for enhancing customer interactions while minimizing customer concerns (Buhalis et al., 2023). This book chapter will focus on how SMEs in the hospitality sector can use smart technologies to enhance customer experiences and interactions to build organizational resilience.

The Context: SMEs in the Hospitality Industry

Small- and medium-sized enterprises (SMEs) in the hospitality industry are pivotal for economic growth, specifically in terms of employment creation (Motta & Sharma, 2020). The importance of SMEs in the hospitality industry (accommodation and food services) in countries across the globe has been highlighted in various regions—namely Latin America (Motta, 2017), Africa (Mage, 2010), and Europe—for example, the UK has been highlighted by recent research that has revealed that 69.8 percent of all businesses in this sector are SMEs (Business News Wales, 2022).

According to Statista (2022), the current market size of the smart hospitality industry worldwide is USD 13.9 billion in 2020 and is forecast to accelerate and reach USD 58.1 billion by 2027. A study by the Connected Commerce Council (2021a) reveals that during the pandemic, digitally advanced SMEs outperformed their counterparts on a range of business outcomes, such as revenue and maintaining customers. Moriarty et al. (2008, p. 309) highlight the significant need for further studies which provide greater clarity on the role of smart technologies. It is assumed that bigger hotels (chains) can be the best candidates for implementing smart technology (Alsetoohy et al., 2019; Koseoglu et al., 2016; Rahimi, 2017). However, smart tech, such as big data and AI, can be beneficial for independent hotels to compete with larger branded hotel chains that have superior budgets, brand salience, and geographic coverage. Such technologies provide extensive insight into the consumer perspective, allowing independent hotels to capitalize on their existing strengths (higher customer engagement and flexibility) to improve their more intimate guest experience advantages (Lee et al., 2020a). Therefore, a comparative analysis of different smart technologies with a focus on how/when they can be beneficial for SMEs in the hospitality sector will be done.

This chapter proposes a model which illustrates how smart technologies can enhance the value creation and value capture aspects of organizations in the hospitality sector, in order to improve the customer experience and build organizational resilience. Furthermore, an examination of how factors such as customer acceptance of technology, willingness to use, privacy/security concerns, and perceived usefulness of smart technology impact customer experiences and the ability of organizations in the hospitality industry to deliver customer value through enhanced interactions will be undertaken (Floros et al., 2021; Han et al., 2021; Yang et al., 2021; Rahimi et al., 2017; Vatankhah et al., 2023). In order to provide insights into how smart technology can enhance organizational resilience in the hospitality sector, the chapter will start with an introduction to the concept of organizational resilience. This will be followed by a discussion of the most common smart technologies used in the hospitality and an analysis of how they can be used by SMEs in the sector. We then introduce the concept of business model innovation and the role of smart technology in it, followed by considerations of smart technology experiences from the customer point of view. Finally, we integrate our discussion into a conceptual model of building organizational resilience through the use of smart technologies.

Organizational Resilience

The topic of resilience has raised several debates in the academic literature, with many definitions used from different research areas (Doorn, 2017). To date, there is no one universally accepted definition for the concept of organizational resilience (Melian-Alzola et al., 2020). The current definitions can be divided into two broad categories: firstly, resistance and recovery from a shock, change, or disaster; and secondly, innovation and adaptation to ensure competitive positioning in the market. Organizational resilience based on the first category is described as the organizational capacity to absorb the impact and recover from an actual occurrence or an extreme weather event, for example, an earthquake (Linnenluecke et al., 2012), or dealing successfully with other types of crises (such as economic/financial crises, geopolitical changes, climate change impacts, as well as internal organizational factors). Furthermore, Boin and Eeten (2013) focused on the element of the organization bouncing back to normal operations after being limited by an event. The second category includes resilience being defined as more than survival (Somers, 2009). It focuses on building capacity (Ortiz-de-Mandojana & Bansal, 2016) to anticipate, identify, and adapt organizational capabilities over time (Reinmoeller & Van Baardwijk, 2005) to strengthen the organization (Vogus & Sutcliffe, 2007). Subsequently, the working definition for organizational resilience here is the ability of an organization to adapt, absorb, anticipate, and develop responses to external events to ensure organizational survival and operational continuity.

The hotel industry is exposed to many external forces (for example, economic crises, climate change, the COVID-19 pandemic) that force organizations to change to stay in business (Melian-Alzola et al., 2020).

The force of the turbulence can differ in size and frequency (Bhamra, 2015). According to Ho et al. (2022) the hotel industry is more susceptible to the effects of disasters than most other industries as the hotel industry requires the free movement of people. According to Ewertowski (2020) there are four key capabilities that an organization needs to focus on developing to improve their response to a crisis: (1) responding, (2) monitoring, (3) anticipating, and (4) learning from the incident. Even with the development of the aforementioned capabilities and crises, unexpected events can happen without notice. With the hotel industry's vulnerability to unexpected events, how to improve and build resilience is a key issue that needs to be investigated (Melian-Alzola et al., 2020). One of the key mechanisms to enable organizations to become more resilient is to integrate and utilize technology (Fulco et al., 2019).

Smart technologies can facilitate the creation of an organizational protective "bubble" that can fortify its capabilities to survive and prosper after sudden changes or turmoil. The key input of smart technologies relates to their manifold synergetic nature that makes possible an optimized, real-time, and focused organizational action both in planning and in execution. Perhaps even more important is the fact that employing smart technologies allows for emerging tactical adjustments in the way crises are handled, according to changes in the type and content of the data flows (see big data) during sudden events.

Smart Technologies

Technology plays a crucial role in enhancing, co-creating, and personalizing the experience of guests (Sharma & Gupta, 2021). The smart concept has come about through the advancement of information technology systems that integrate hardware, software, and network technologies together to help people make more informed decisions (Abdelmoaty & Soliman, 2020). Smart technologies can be conceptualized as an information and communication mechanical system that is equipped with sensors, actuators, and pre-programmed controllers which allow a structure to adapt to unpredictable external condition to improve the management business processes (Holnicki-Szulc, 2008). That means that smart technologies enable machines to become aware of the real world and data around them in real time which allows them to act on their own or help people in making better business decisions (Gretzel et al., 2015). A variety of different smart technologies have been reported to be utilized and diffused in the hospitality industry, including mobile apps/platforms; artificial intelligence; big data, and robotics. These four smart technologies have started becoming particularly popular in the tourism sector as they practically link to various service settings by advancing their value capturing, creation, and diffusion. Therefore, we focus on those technologies here.

The Role of Mobile Apps/Platforms, AI, Robotics, and Big Data for SMEs in the Hospitality Industry

Mobile apps/platforms primarily provide smartphone users with hotel and travel information. Apps and platforms such as Expedia and Booking.com can all be accessed from a person's handheld device to gain access to offers, products, views, and rating in order to contribute to the decision-making process (Abdelmoaty & Soliman, 2020). Mobile apps allow for a more personalized experience for users and information personalization has a significant and positive effect on perceived benefits associated with mobile app use increasing booking intentions for hotels (Cobos & Shapoval, 2022). Research on VR/AR mobile apps suggests that they add both extrinsic value (efficiency and better shopping) and intrinsic value (entertainment) while also increasing conversion rates and improved customer satisfaction (Dacko, 2017). A concern with mobile apps is the number of apps a user has on their smartphone; for instance, respondents from a study conducted in China noted that it was troublesome to download an app for a one-time visit, but others found the convenience outweighed the annoyance of another app (Liu & Hung, 2020). Moving on from web 2.0 social media platforms to web 3.0 new technologies, artificial intelligence is one of the most promising technological areas for the future.

Artificial intelligence (AI) can be seen as intelligent interfaces (e.g., avatars, metaverse) and advanced software (e.g., voice assistants) with a degree of autonomy and possibly mobility and sensory capabilities that allow them to perform tasks without human intervention (Popesku & Zaitanov, 2019). AI can provide advantages including cost reduction, revenue generation, and improved customer experience. Electricity usage can be managed automatically by AI-based electricity back-up systems, chemical management systems, and motion detection systems to reduce costs; alternatively, AI can also generate revenue through improved up-selling and marketing advantages (Nam et al., 2021). However, AI related to food waste management is unlikely to be accepted by managers of small hotels due to slow ROI, challenges with system integration, threat of imminent changes/improvements, and environmental uncertainty (Alsetoohy et al., 2019). A special type of AI is robotics as they embody many advanced intelligent features in the form of humanoids.

Robotics, as specific form of AI, have a high initial investment and continuing maintenance costs, so small and medium businesses would possibly adopt this smart technology in cases that cost/benefit analysis would indicate use of service robots offers significant benefits; thus, it is expected that once the cost/benefit is acceptable, there will be a rapid adoption of robotics (Kuo et al., 2017). COVID-19 improved the perceived usefulness of robots in hotels—making people more willing to use robots, resulting in a preference for robot service compared to human service, which is different from most previous studies conducted before the COVID-19 pandemic, which indicated a preference for human service rather than robot service in hotels (Kim et al., 2021a). It has been found that the degree of acceptance of robots in hospitality settings is strongly linked to the degree of acceptance based on travelers' nationalities (Mukherjee et al., 2021). According to Mukherjee et al. (2021), Chinese travelers have been found to be the most accepting while German travelers the least. It is currently unlikely for small- and medium-sized hotels to benefit from the implementation of robotics; however, depending on the rate of improvement in both function and affordability, robotics could be beneficial in the future. Various forms of artificial intelligence, including robotics, function on the basis of complex algorithms and big data databases, collected from various sources.

Big data can be defined as large amounts of different types of data produced from various types of sources, such as people, machines, or sensors (The EU Data Protection Reform and Big Data, 2016). Big data can be used to combine sales data, customer reviews, social media postings, users' social network, and activity information to understand hotel guest experience (Lee et al., 2019). Big data can assess customer satisfaction through data mining and text mining, allowing management to make a timely response to both positive and negative reviews and allow for customer satisfaction-oriented strategy development (Lee et al., 2020a; Vassiliadis & Stylos, 2022). Specifically, social media analytics can help strengthen the decision-making processes through information provided by comment analysis, message frequency analysis, and brand purchase frequency analysis (Zhang et al., 2022). Big data can provide access to an abundance of relevant information to hotel management; however, this smart technology is only beneficial if the information is used effectively to improve internal operations and the guest experience (Yallop & Seraphin, 2020).

Comparative Analysis of the Chosen Smart Technologies

In the section to follow the smart technologies discussed in the previous section are tabled based on the value they offer to business (Table 4.1) and the customer (Table 4.2).

The preceding section demonstrates how smart technologies can be used to augment both the SMEs' processes and consumers' experiences in the hospitality industry. We now turn to a discussion on business model innovation with a view of how smart technologies can enhance the value creation and value capture mechanisms for SMEs.

Business Model Innovation

The concept of a business model (BM) has gained prominence in the management literature along with a focus on business model innovation (Foss & Saebi, 2017) to demonstrate how business processes and features need to be considered. A business model is defined as the "design or

Table 4.1 Business viewpoint of the value of smart technologies (Source: Authors)

Value to business	Mobile apps/ platforms	AI-powered interfaces	Robotics	Big data
Increased visibility	x	x		
Personalized experience for customer	x	x		x
Increased booking intention	x			
Increased automation		x	x	
Reduction in operational costs (e.g., electricity usage)		x	x	
Personalized marketing		x		x
Increased operating effectiveness		x		x
Enhanced product and market strategy				x
Enhanced planning and decision making tool		x		x

Table 4.2 Customer viewpoint of the value of smart technologies (Source: Authors)

Customer viewpoint	Mobile apps/ platforms	AI-powered interfaces	Robotics	Big data
Increased access to business offerings	x			
Access to hotel ratings for decision making	x			x
Removal of language barriers	x	x		
Cost reductions (Loyalty, agility)	x		x	x
Bespoke offerings		x		x
Reduced human interactions (e.g., COVID-19)		x	x	
Streamlined processes improving overall experience		x		x

architecture of the value creation, delivery, and capture mechanisms" of an organization (Teece, 2010, p. 172). BM has been used in the extant literature to describe how an organization creates and delivers value to customers, as well as how the organization captures (or monetizes) value in return (Zott et al., 2011). A well-designed business model is crucial for an organization's long-term sustainability. Importantly, business models need to be flexible enough to adapt to changes in technology, organizational innovation, or external environmental changes (Teece, 2010) to be relevant and appropriate.

Business model innovation (BMI) is referred to as "the process of changing and reconfiguring the building blocks that define the structure of an existing business model" (Reficco et al., 2021, p. 721). BMI can affect changes to the different business model components, such as the value proposition and value creation of the organization (products, services, core activities, and processes), as well as the value capture mechanisms (e.g., revenue generation, operational efficiencies) (Amit & Zott, 2012; Bocken et al., 2014; Weerawardena et al., 2020; Buhalis et al., 2023; Gannon et al., 2022; Bihamta et al., 2017). BMI can be particularly important for the resilience of SMEs under rapid technological and environmental change. A recent survey of European SMEs found that business model innovation and novel business models created lasting performance benefits, highlighting that early market entrants have more experienced management which, in combination with the flexibility of SMEs, creates opportunities to foster BMI, thus making firms more responsive to technological developments (Molina-Castillo et al., 2022).

Using smart technologies as part of the process of BMI has been found to have a positive influence on the competitive advantage indicators, such as profit, sales, ROI, and market share (Dymitrowski & Mielcarek, 2021). Furthermore, innovation has been helpful to SMEs to mitigate the effects of uncertainty (Verreynne et al., 2019). In hospitality, the areas of value creation include guest experience, operational efficiency, and environmental sustainability (Nadkarni et al., 2020). The traditional hotel business model is aging, and hotels are being required to adapt their technologies to remain competitive. For instance, user-generated content (UGC) is a key source for creating big data inputs and can support various mobile platforms and AI-powered interfaces in producing well-informed outputs to the benefit of the tourists/guests and the organizations in the tourism sector (Stylos et al., 2021b, b). UGC offers free-will/voluntary information, and unbiased inputs on behalf of tourism consumers, especially during times of volatility in service settings, as it facilitates synergies, collective decision-making, and optimized outputs for all and each one of the stakeholders in the tourism business ecosystem. Additionally, big data can be used to analyze social media and improve decision-making processes to create a new "data-assisted social media analytics for business model (BD-SMAB)" (Zhang et al., 2022).

When there is industry turbulence, like a disruptive natural event, businesses are likely to gain a sustainable competitive advantage through business model innovation. By adopting the new technological advancements, organizations are able to introduce innovation in marketing (e.g., promotion, pricing, channels, and packaging) to respond to the uncertainties of the competitive landscape (Presenza et al., 2019). By innovating, these organizations are able to adapt their offerings using, for example, smart technologies and, thus, being more resilient to these uncertainties in the environment. It should be noted that a recent study explored the possible negative impacts of BMI on customers—how change can prompt negative responses depending on their individual cognitive style and satisfaction with the established BM; however, this study suggests ways to mitigate negative responses including simplifying the change, highlighting benefits of the new BM, and a slow roll-out to allow customer feedback (Heidenreich et al., 2022).

Using Smart Technology to Enhance Customer Experiences and Interactions

With regard to customer experiences and interaction, an important aspect to consider is the ability of technology to allow co-creation opportunities for customers. Research suggests that co-creation innovation via smartphones increases interaction and involvement which is suggested to increase the consumers' willingness to adopt, which could be beneficial for smart technology (Sarmah et al., 2017). However, an empirical study in China tested co-creation in terms of customer involvement in new hotel service development and concluded that increasing consumer involvement does not necessarily improve the relationship between business and consumer when staff and resources are inadequate (Ma et al., 2017). For SMEs that are often time resource constrained, it is important to consider and carefully balance the implementation of technology with other resources and employees' availability and training. Customer experiences and interactions need to be considered holistically, where technology is only one part of the value (co)creation system.

Considerations related to customer preferences and SMEs' ability to achieve ROI need to be taken into account. For example, a study that segments the market into perceptions/attitudes toward technology instead of type of traveler/hotel user preference suggested that only individuals that enjoy technology are willing to pay more for in-room smart technology (Erdem et al., 2019). Similarly, in a study on unmanned smart hotels, when guests experience high-quality stays, motivation, confidence, satisfaction, and loyalty are positively impacted, resulting in a continued high degree of interest in unmanned smart hotels, suggesting that positive experiences with smart technology encourages overall acceptance (Chang et al., 2022). Additionally, research suggests that the acceptance of smart technology in the industry can be incentivized further by showing that the technology is beneficial for both the hotel (save costs) and customers (avoid poor service), to avoid being viewed as only a cost-saving strategy (Belanche et al., 2020). Using smart technology to enhance customer experiences in the hospitality industry requires consideration of two critical aspects: SMEs' ability to use technology and customers' perceptions about smart technology. We discuss those in turn.

SMEs' Ability to Use Smart Technology

SMEs have easier access to the applications of the Fourth Industrial revolution, as this is data-driven rather than heavy infrastructure based. The democratization effects of digital technologies are said to be especially important to small entrepreneurial firms, which are usually resource-constrained (Nambisan, 2017). Thus, enhanced flexibility to achieve business transformation and equal opportunities for access to this new industrial economy can improve their business competitiveness. Additionally, this could help SMEs toward building high levels of resilience during crises such as COVID-19.

Notwithstanding the potential of digital technologies to help SMEs across the service sectors to recover from the pandemic, there are several potential *barriers* that need to be understood in order for those organizations to fully take advantage of the affordances presented by smart technology. A Bank of Canada report, focused on the service sector, has documented that "firms view digital technologies as a driver of actual or expected changes in market structure, citing consolidation and concentration of market power among dominant players, forcing smaller players out of the market" (Dong et al., 2017, p. 3). Thus, although smart technologies bring the potential for democratization and improved performance, the realization of this potential depends on the ability of SMEs to deploy technology solutions effectively. Among the major barriers for smart technology implementation by SMEs are cognitive barriers, such as reluctance to change, and worries over the costs/ROI from smart technology (Dong et al., 2017; Connected Commerce Council, 2021b; Molina-Castillo et al., 2023; Rezaei et al., 2018; Sigala et al., 2019; Rahimi et al., 2023). Within the hospitality industry, major barriers to implementing smart hotel technologies are: technology barriers (e.g., the extent of changes in the infrastructure necessary for implementing smart technologies) (Mariani & Wirtz, 2023), financial barriers (e.g., concerns about the benefits from investing in smart technologies), attitude and behavioral barriers (e.g., reluctance to change), and environmental barriers (e.g., privacy and security legislation, stakeholder concerns about data mining) (Stylos et al., 2021b, b).

SMEs in the hospitality industry, thus, present an interesting paradox: on the one hand, the fourth industrial revolution offers them the affordances of smart technologies that can help them recover from the COVID-19 crisis; on the other hand, numerous barriers to implementation of smart technologies exist, and worries over market consolidation that forces smaller players out persist. While recent academic research has highlighted the importance of digital technologies for SMEs' survival during and post-pandemic, as well as for business model innovation (e.g., Priyono et al., 2020), the priority of innovation models tends to focus on larger organizations (e.g., Urbaniec & Żur, 2021) or new innovative (high-tech) start-ups (Foss & Saebi, 2017).

We know from the digital entrepreneurship literature that digital technologies profoundly affect the abilities of small entrepreneurial ventures across various sectors to operate and grow (Nambisan, 2017). However, there is a dearth of models explaining how smart technologies are applied by SMEs for achieving resilience, particularly in volatile sectors profoundly impacted by crises such as the COVID-19. The digital transformation of small service firms in the hospitality sector affects various aspects of *service delivery* and performance, with a focus on the *design of new business models*, and the ability of SMEs to build resilience post-pandemic. Since business models are directly related to the creation, delivery, and capture of value (Teece, 2010), managers need to master the ability to recognize when innovation in their business model is needed.

Smart Technology and Customer Perceptions in the Hospitality Industry

Customer perceptions about smart technology in the hospitality industry can vary depending on consumers' demographic and cultural characteristics. Variables such as consumer age and culture have been found to affect perceptions about usefulness and benefits of smart technology. Because consumers' perceptions are a key factor in consumers' willingness and ability to co-create value using the available technology, they need to be included into a business model innovation framework. For instance, a study in the UK including interviews with millennials concluded that they had "negative perceptions of daily technology use" and "perceived benefits to digital-free travel" which suggests that millennials may desire an escape from technology when on vacation (Floros et al., 2021). However, the opposite was found in a Chinese study that determined that young adults prefer smart hotels and enjoy the challenge/unique experience of smart technology in hotels (Yang et al., 2021). Smart technology that requires the guest to have an updated smartphone raises concerns about consumers with older/outdated phones being unable to access the mobile technology; therefore, investing in mobile smart tech will only be beneficial if the main guest segment owns and uses updated smartphones (Nam et al., 2021). A South Korean study determined the effect of entertainment on attitude was stronger for older generations, suggesting that in South Korea older people find smart technology more impressive, entertaining, and exciting (Kim et al., 2021a). However, the opposite was found in a European study that noted older individuals had concerns with the overlap of technology and tourism because the older age group perceived an excessive use of technology by young users and warned about the impact this can have on privacy, socialization processes, and mental health (Femenia-Serra et al., 2022). A cross-country comparison between the United States and South Korea highlights that there is a cultural difference in the internal inclination of travelers to adopt smart tourism technologies; therefore, suggesting that hotels need to adjust the technology depending on their guest demographics and their target market segments (Shin et al., 2021).

Keeping those considerations in mind, SMEs in the hospitality industry need to understand how customer perceptions will affect the value creation process in their efforts to innovate their business model through smart technology (Stylos & Vassiliadis, 2023). Specifically, we propose that perceived security, perceived usefulness and willingness to use, and personalization will ultimately affect customer acceptance, and will be key factors that will determine the extent to which implementation of smart technology can contribute to customer experiences, and subsequently to the long-term viability and resilience of the business. We discuss each of those factors in turn.

Perceived Security

Perceived security has a positive impact on consumers' intention to use smart technology when the smart technology is also compatible with a guest's existing beliefs/values and experiences (Han et al., 2021). A study assessing hotel customers' intention to use fingerprint systems suggested that fingerprint systems can be perceived as useful to improve convenience and security levels, but privacy concerns were apparent; specifically, personal concerns from older consumers lowered perceived usefulness and, in turn, lowered intention to use (Kim & Bernhard, 2014). A European study highlighted the variability of privacy concerns and concluded that tourists often do not trust most organizations with potential access to their personal data and lack specific knowledge of data management policies and practices (Femenia-Serra et al., 2022). However, despite privacy concerns, self-confident individuals still adopted smart technology due to its perceived value, but privacy risk negatively affected adoption intention for the individuals less familiar with smart tech, suggesting that more user experience might reduce privacy concerns against the adoption intention (Cao et al., 2022).

Perceived Usefulness and Willingness to Use

A South Korean study on the technology readiness of consumers confirmed that both perceived ease of use and perceived usefulness influenced attitudes which in turn generated favorable behavioral intentions (Kim & Han, 2022). If the smart hotel technology is easy to use, customers are more likely to perceive the technology as useful (Yang et al., 2021). COVID-19 improved the perceived usefulness of robots in hotels—making people more willing to use robots, resulting in a preference for robot service compared to human service (Kim et al., 2021a). The previous study indicates that guests are more willing to use smart technology when concerns around health and safety are more apparent. A study on Chinese young adults and early middle-aged adults suggests that smart technology in hotels that is more difficult to use actually increases visit intention—which the researchers suggested is likely due to the higher acceptance, increased interest, and overall willingness to use smart technologies (Yang et al., 2021).

Personalization

Personalization can increase customers' perceived usefulness and willingness to use smart technology. A study investigating smart technology in luxury hotels in China stated that smart technology provides standardized services for consistency but lacks personalization because of limited functionalities (Liu & Hung, 2020). Personalization is noted as a main advantage of a luxury hotel experience; therefore, the customer experience could be damaged if the technology lacks personalization capabilities (Liu & Hung, 2020). The personalization capabilities of mobile apps facilitate the communication of perceived benefits (leaving positive reviews), reduces perceived risk, increases trust in the hotel's health-related messages, and reduces any health-related worries, resulting in increased booking intentions (Cobos & Shapoval, 2022). A Korean study that investigated the impact of age and gender on consumer behavior concluded that perceived value is affected by personalization and entertainment for all groups (Kim et al., 2021b). A more specific example of how to increase a guest's willingness to use smart technology is to use personalized avatars based on the socio-demographic profile of the guest because people prefer avatars that look like them, resulting in perceived expertise and increased intention to use (Choi et al., 2020).

Discussion

Model/Framework of Resilience

Building on the discussion in the preceding sections, we offer a model of BMI in the hospitality industry based on smart technology use. The model depicts the parts for the business model innovation along with the main constructs considered important in the context of digital transformation for SMEs in the hospitality sector. The outcome focuses on the *resilience* of these organizations and their use of smart technologies. The initial inputs focus on the types of technologies required for the process of *value (co)creation*. In this context, the BMI is based on the different business model components, such as the value proposition and value creation of the organization (products, services, core activities, and processes), as well as the value capture mechanisms (e.g., revenue generation, operational efficiencies) (Amit & Zott, 2012).

The SME's ability to *use smart technologies* is important in the context of creating smart technology value propositions and *organizational resilience* (Azis et al., 2020). For the model, this means that the technology is useful if customers are able to use it; hence, it requires consideration of the key abilities needed for different customers. Similarly, customer *usefulness of technology and security* is seen as an important factor contributing to the *SME organizational resilience* (Floros et al., 2021; Han et al.,

2021; Kim & Bernhard, 2014; Yang et al., 2021). These are important contributors to the *value (co)creation and delivery* connected to the BMI perspective.

The combinations of different *smart technologies* used along with a blueprint of the *customer experiences and interactions* can significantly enhance the organizational resilience for SME organizations in the hospitality sector (Ma et al., 2017). As per various research projects on the important role of the customer experiences and stakeholder interactions for informing business model innovation, the organizational resilience can be upgraded through better evaluation of resources, targets toward decision-making optimization, when smart technologies are used (Ma et al., 2017; Mariani et al., 2021; Sarmah et al., 2017). This is dependent on the usefulness of *customer experiences and interactions* taken from direct feedback and also captured from indirect feedback such as from data about experiences such as booking information, social media feedback, etc. The content generated by the customer such as user-generated content (UGC) is also of great value in addressing organizational resilience (Fig. 4.1).

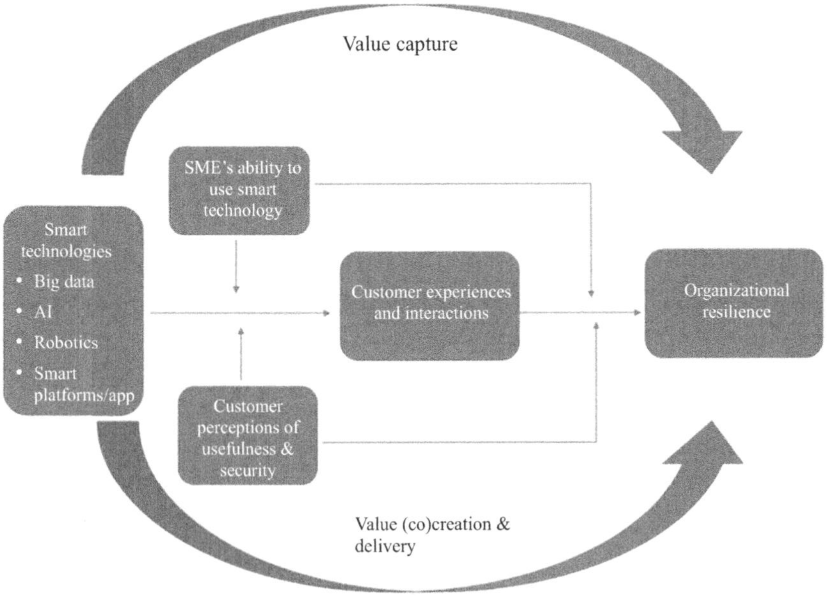

Fig. 4.1 Smart technologies' contribution to SME organizational resilience (Source: Authors)

Practical Implications

The practical aspect of employing smart technologies in SMEs of the tourism and hospitality sector with a particular focus on enhancing organizational resilience has been of particular interest, especially due to the crises that have taken place in the last decade (i.e., financial, health-related/pandemic, and wars). There is need for all tourism organizations, particularly SMEs, to be equipped with abilities to forecast, prepare for, and adequately respond to different types of crises (external or internal causes) that would increase the possibilities of survival or even prosperous business venturing in extreme circumstances. Maintaining a close relationship with customers (tourists, guests, travelers), understanding their real-time needs, and any possible changes in their attitudes and decision-making patterns, to ultimately create value as circumstances change can be better supported through sophisticated smart platforms. These platforms can function as intelligent hubs that would help all stakeholders make the best possible informed and real-time decisions. These decisions may relate to selecting tourism destinations, making travel arrangements, booking accommodations, and determining other services related to the tourist experience. By engaging tourists in the smart platforms, tourism organizations gain access to information relating to their desires, specific needs, perceived constraints, and any other parameters influencing tourists' decision-making mechanisms.

Because of the resource constraints of SMEs, another aspect to consider is the place of the organization in the tourism ecosystem, and the extent to which SMEs can leverage common resources. For example, smart tourist destinations use smart technology to improve tourist experience making the destination as a whole more competitive/appealing compared to surrounding destinations (Cimbaljevic et al., 2019). Smart technologies can make a tourism experience more memorable, leading to increased tourist satisfaction, which increases destination loyalty; therefore, smart technology in a hotel can improve a tourist destination as a whole (Azis et al., 2020). However, government involvement is required in the process of creating smart tourism cities to ensure the local residents are not negatively impacted by increased tourism while still creating a unique positive experience for tourists (Lee et al., 2020b). Big data analytics can contribute to the destination management of an entire tourism industry of a location, not just one hotel; suggesting that big data can be used throughout the value chain with different stakeholders to benefit a tourist destination by helping managers and marketers (Mariani et al., 2021; Stylos & Zwiegelaar, 2019). Notably, the emerging sharing economy, Xbed and Airbnb, in the hotel industry poses a threat to traditional hotels, which could suggest the importance of smart technology to differentiate and reduce costs (Leung et al., 2019).

Conclusion and Recommendations

Key benefits can be expected from employing smart technologies in SMEs not only at tactical and operational levels but also at strategic levels. In this regard, the organizational structures can become more flexible in responding to real-time service requests, better informed in their decision-making, better positioned in competing with other organizations, and better equipped in facing challenges and risks arising from the external environment. In essence, it is the value capture, creation, and delivery chain that makes a difference for capitalizing the smart technologies benefits in the form of building stronger organizational resilience. This is extremely important in the tourism industry which is dominated by SMEs who experience great challenges in the contemporary volatile service environment. The conceptual model proposed in this chapter offers a holistic view of the key aspects to be considered both from a consumer/hotel guest point of view and from the SME perspective, when implementing smart technologies in the hospitality industry.

Recommended Further Reading

He, Z., Huang, H., Choi, H., & Bilgihan, A. (2021). Building organizational resilience with digital transformation. *Journal of Service Management, 34*(1), 147–171. https://doi.org/10.1108/JOSM-06-2021-0216

Schwaiger, K., Zehrer, A., & Braun, B. (2022). Organizational resilience in hospitality family businesses during the COVID-19 pandemic: A qualitative approach. *Tourism Review, 77*(1), 163–176. https://doi.org/10.1108/TR-01-2021-0035

Senbeto, D., & Hon, A. (2020). Market turbulence and service innovation in hospitality: Examining the underlying mechanisms of employee and organizational resilience. *The Service Industries Journal, 40*, 1119–1139. https://doi.org/10.1080/02642069.2020.1734573

Stylos, N. (2022). An integrated duality theory framework (IDTF): Marking pathways for consumer decision-making researchers in the hospitality and tourism industry. *International Journal of Contemporary Hospitality Management, 34*(7), 2597–2619.

Su, D., Tra, D., Huynh, H., Nguyn, H., & O'Mahony, B. (2021). Enhancing resilience in the Covid-19 crisis: Lessons from human resource management practices in Vietnam. *Current Issues in Tourism, 24*(22), 3189–3205. https://doi.org/10.1080/13683500.2020.1863930

Troisi, O., Visvizi, A., & Grimaldi, M. (2023). Digitalizing business models in hospitality ecosystems: Toward data-driven innovation. *European Journal of Innovation Management, 26*(7), 242–277.

Xie, X., Wu, Y., Palacios-Margues, D., & Ribeiro-Navarrete, S. (2022). Business networks and organizational resilience capacity in the digital age during COVID-19: A perspective utilizing organizational information processing theory. *Technological Forecasting and Social Change, 177*. https://doi.org/10.1016/j.techfore.2022.121548

References

Abdelmoaty, G. A., & Soliman, S. A. E. M. (2020). Smart technology applications in tourism and hospitality industry of the new administrative capital, Egypt. *Journal of Association of Arab Universities for Tourism and Hospitality, 19*(2), 102–129. https://doi.org/10.21608/JAAUTH.2020.47549.1085

Alsetoohy, O., Ayoun, B., Arous, S., Megahed, F., & Nabil, G. (2019). Intelligent agent technology: What affects its adoption in hotel food supply chain management? *Journal of Hospitality and Tourism, 10*(3), 286–310. https://doi.org/10.1108/JHTT-01-2018-0005

Amit, R., & Zott, C. (2012). Creating value through business model innovation. *MIT Sloan Management Review, 53*(3), 41–49. https://doi.org/10.1108/JHTT-01-2020-0005

Azis, N., Amin, M., Chan, S., & Aprilia, C. (2020). How smart tourism technologies affect tourist destination loyalty. *Journal of Hospitality and Tourism Technology, 11*(4), 603–625. https://doi.org/10.1108/JHTT-01-2020-0005

Belanche, D., Casalo, L., & Flavian, C. (2020). Frontline robots in tourism and hospitality: Service enhancement or cost reduction? *Electronic Markets, 31*, 477–492. https://doi.org/10.1007/s12525-020-00432-5

Bhamra, R. (Ed.). (2015). *Organizational resilience: Concepts, integration, and practice*. CRC Press.

Bihamta, H., Jayashree, S., Rezaei, S., Okumus, F., & Rahimi, R. (2017). Dual pillars of hotel restaurant food quality satisfaction and brand loyalty. *British Food Journal, 119*(12), 2597–2609.

Bocken, N. M. P., Short, S. W., Rana, P., & Evans, S. A. (2014). A literature and practice review to develop sustainable business model archetypes. *Journal of Cleaner Production, 2014*(65), 42–56. https://doi.org/10.1016/j.jclepro.2013.11.039

Boin, A., & Van Eeten, M. J. (2013). The resilient organization. *Public Management Review, 15*(3), 429–445. https://doi.org/10.1080/14719037.2013.769856

Buhalis, D., O'Connor, P., & Leung, R. (2023). Smart hospitality: From smart cities and smart tourism towards agile business ecosystems in networked destinations. *International Journal of Contemporary Hospitality Management, 35*(1), 369–393.

Business News Wales. (2022). *New analysis reveals the importance of SMEs to the British Economy*. https://businessnewswales.com/new-analysis-reveals-the-importance-of-smes-to-the-british-economy-4/

Cao, D., Sun, Y., Goh, E., Wang, R., & Kuiavska, K. (2022). Adoption of smart voice assistants' technology among Airbnb guests: A revised self-efficacy-based value adoption model (SVAM). *International Journal of Hospitality Management, 101*. https://doi.org/10.1016/j.ijhm.2021.103124

Chan, E. S., Okumus, F., & Chan, W. (2018). Barriers to environmental technology adoption in hotels. *Journal of Hospitality and Tourism Research, 42*(5), 829–852. https://doi.org/10.1177/1096348015614959

Chang, Y., Cheah, J., Lim, X., Morrison, A., & Kennell, J. (2022). Are unmanned smart hotels du jouror are they here forever? Experiential pathway analysis of antecedents of satisfaction and loyalty. *International Journal of Hospitality Management, 104*. https://doi.org/10.1016/j.ijhm.2022.103249

Choi, Y., Mehraliyev, F., & Kim, S. (2020). Role of virtual avatars in digitalized hotel service. *International Journal of Contemporary Hospitality Management, 32*(3), 977–997. https://doi.org/10.1108/IJCHM-03-2019-0265

Cimbaljevic, M., Stankov, U., & Pavlukovic, V. (2019). Going beyond the traditional destination competitiveness – Reflections on a smart destination in the current research. *Current Issues in Tourism, 22*(20), 2472–2477. https://doi.org/10.1080/13683500.2018.1529149

Cobos, L., & Shapoval, V. (2022). Leveraging psychological characteristics to influence mobile hotel bookings during a global health crisis. *International Journal of Hospitality & Tourism Administration, 1*(24). https://doi.org/10.1080/15256480.2022.2103485

Connected Commerce Council. (2021a). *Digitally driven / Europe: Travel. European small and medium sized enterprises (SMEs).* Transformation, Innovation, and Resilience During the COVID-19 Pandemic. https://bit.ly/3fmrhLb

Connected Commerce Council. (2021b). *Digitally driven / Europe. European small and medium sized enterprises (SMEs).* Transformation, Innovation, and Resilience During the COVID-19 Pandemic. https://digitallydriven.connectedcouncil.org/europe/

Dacko, S. (2017). Enabling smart retail settings via mobile augmented reality shopping apps. *Technological Forecasting and Social Change, 124*, 243–256. https://doi.org/10.1016/j.techfore.2016.09.032

Dong, W., Fudurich, J., & Suchanek, L. (2017). *Digital transformation in the service sector: Insights from consultations with firms in wholesale, retail and logistics.* https://www.bankofcanada.ca/2017/11/staff-analytical-note-2017-19/

Doorn, N. (2017). Resilience indicators: Opportunities for including distributive justice concerns in disaster management. *Journal of Risk Research, 20*(6), 711–731. https://doi.org/10.1080/13669877.2015.1100662

Dymitrowski, A., & Mielcarek, P. (2021). Business model innovation based on new technologies and its influence on a company's competitive advantage. *Journal of Theoretical and Applied Electronic Commerce Research, 16*(6), 2110–2128. https://doi.org/10.3390/jtaer16060118

El-Said, O. A., Smith, M., Al-Yafaei, Y., & Abdul Salam, S. A. (2023). From complexity to evolution: Mapping resilience management practices in the hospitality industry during the COVID-19 crisis. *International Journal of Hospitality Management, 110*, 103435. https://doi.org/10.1016/j.ijhm.2023.103435

Erdem, M., Atadil, H. A., & Nasoz, P. (2019). Leveraging guest-room technology: A tale of two guest profiles. *Journal of Hospitality and Tourism Technology, 10*(3), 255–268. https://doi.org/10.1108/JHTT-06-2018-0046

Ewertowski, T. (2020, November). An organizational resilience potential scale within the hospitality industry in the face of the COVID-19 in Poland. In *IBIMA conference*. https://ibima.org/accepted-paper/an-organizational-resilience-potential-scale-within-the-hospitality-industry-in-the-face-of-the-covid-19-in-poland/

Femenia-Serra, F., Ioannou, A., & Tussyadiah, I. (2022). Is smart scary? A mixed-methods study on privacy in smart tourism. *Current Issues in Tourism, 25*(14), 2212–2238. https://doi.org/10.1080/13683500.2021.1987399

Floros, C., Cai, W., McKenna, B., & Ajeeb, D. (2021). Imagine being off-the-grid: Millennials' perceptions of digital-free travel. *Journal of Sustainable Tourism, 29*(5), 751–766. https://doi.org/10.1080/09669582.2019.1675676

Foss, N. J., & Saebi, T. (2017). Fifteen years of research on business model innovation: How far have we come, and where should we go? *Journal of Management, 43*(1), 200–227. https://doi.org/10.1177/014920631667592

Fulco, I., La Sala, A., & Loia, F. (2019). The role of technology for building resilience thinking in corporate governance. *New Challenges in Corporate Governance: Theory and Practice*, 206–219. https://doi.org/10.22495/ncpr_36

Gannon, M., Taheri, B., Thompson, J., Rahimi, R., & Okumus, B. (2022). Investigating the effects of service recovery strategies on consumer forgiveness and post-trust in the food delivery sector. *International Journal of Hospitality Management, 107*, 103341.

Gretzel, U., Koo, C., Sigala, M., & Xiang, Z. (2015). Special issue on smart tourism: Convergence of information technologies, experiences, and theories. *Electronic Markets, 25*, 175–177. https://doi.org/10.1007/s12525-015-0194-x

Han, D., Hou, H., Wu, H., & Lai, J. (2021). Modelling tourists' acceptance of hotel experience-enhancement smart technologies. *Sustainability, 13*(8). https://doi.org/10.3390/su13084462

Heidenreich, S., Freisinger, E., & Landau, C. (2022). The dark side of business model innovation: An empirical investigation into the evolvement of customer resistance and the effectiveness of potential countermeasures. *Journal of Product Innovation Management, 39*(6), 824–846. https://doi.org/10.1111/jpim.12627

Ho, G. K., Lam, C., & Law, R. (2022). Conceptual framework of strategic leadership and organizational resilience for the hospitality and tourism industry for coping with environmental uncertainty. *Journal of Hospitality and Tourism Insights*. https://doi.org/10.1108/JHTI-09-2021-0242

Holnicki-Szulc, J. (Ed.). (2008). *Smart technologies for safety engineering*. John Wiley & Sons.

Jiang, Y., & Stylos, N. (2021). Triggers of consumers' enhanced digital engagement and the role of digital technologies in transforming the retail ecosystem during COVID-19 pandemic. *Technological Forecasting and Social Change, 172*. https://doi.org/10.1016/j.techfore.2021.121029

Kim, J., & Bernhard, B. (2014). Factors influencing hotel customers' intention to use a fingerprint system. *Journal of Hospitality and Tourism Technology, 5*(2), 98–125. https://doi.org/10.1108/JHTT-11-2013-0031

Kim, J., & Han, H. (2022). Hotel service innovation with smart technologies: Exploring consumers' readiness and behaviors. *Sustainability, 14*(10), 5746. https://doi.org/10.3390/su14105746

Kim, S., Kim, J., Badu-Baiden, F., Giroux, M., & Choi, Y. (2021). Preference for robot service or human service in hotels? Impacts of the COVID-19 pandemic. *International Journal of Hospitality Management, 93*. https://doi.org/10.1016/j.ijhm.2020.102795

Kim, J., Montes, A., & Han, H. (2021). The role of expected benefits towards smart hotels in shaping customer behavior: Comparison by age and gender. *Sustainability, 13*(4), 1698. https://doi.org/10.3390/su13041698

Koseoglu, M. A., Rahimi, R., Okumus, F., & Liu, J. (2016). Bibliometric studies in tourism. *Annals of Tourism Research, 61*, 180–198.

Kuo, C. M., Chen, L. C., & Tseng, C. Y. (2017). Investigating an innovative service with hospitality robots. *International Journal of Contemporary Hospitality Management*. https://doi.org/10.1108/ijchm-08-2015-0414

Law, R., Ye, H., & Chan, I. C. C. (2021). A critical review of smart hospitality and tourism research. *International Journal of Contemporary Hospitality Management, 34*(2), 623–641. https://doi.org/10.1108/ijchm-08-2021-0986

Lee, M., Cai, Y., DeFranco, A., & Lee, J. (2020). Exploring influential factors affecting guest satisfaction: Big data and business analytics in consumer-generated reviews. *Journal of Hospitality and Tourism Technology, 11*(1), 137–153. https://doi.org/10.1108/JHTT-07-2018-0054

Lee, P., Hunter, W., & Chung, N. (2020). Smart tourism city: Developments and transformations. *Sustainability, 12*(10), 3958. https://doi.org/10.3390/su12103958

Lee, M., Lee, S., & Koh, Y. (2019). Multisensory experience for enhancing hotel guestexperience: Empirical evidence from big data analytics. *International Journal of Contemporary Hospitality Management, 31*(11), 4313–4337. https://doi.org/10.1108/IJCHM-03-2018-0263

Leung, X., Xue, L., & Wen, H. (2019). Framing the sharing economy: Toward a sustainable ecosystem. *Tourism Management, 71*, 44–53. https://doi.org/10.1016/j.tourman.2018.09.021

Linnenluecke, M. K., Griffiths, A., & Winn, M. (2012). Extreme weather events and the critical importance of anticipatory adaptation and organizational resilience in responding to impacts. *Business Strategy and the Environment, 21*(1), 17–32. https://doi.org/10.1002/bse.708

Liu, C., & Hung, K. (2020). A comparative study of self-service technology with service employees: A qualitative analysis of hotels in China. *Information Technology & Tourism, 22*, 33–52. https://doi.org/10.1007/s40558-020-00167-1

Ma, S., Gu, H., Wang, Y., & Hampson, D. P. (2017). Opportunities and challenges of value co-creation: The role of customer involvement in hotel service development. *International Journal of Contemporary Hospitality Management, 29*(12), 3023–3043. https://doi.org/10.1108/IJCHM-08-2016-0479

Mage, G. N. (2010, August). *Financial reporting for small and medium scale enterprises (SMEs) in the hospitality industry*. In *International research symposium in service management*.

Mariani, M., Bresciani, S., & Dagnino, G. (2021). The competitive productivity (CP) of tourism destinations: An integrative conceptual framework and a reflection on big data and analytics. *International Journal of Contemporary Hospitality Management, 33*(9), 2970–3002. https://doi.org/10.1108/IJCHM-09-2020-1102

Mariani, M., & Wirtz, J. (2023). A critical reflection on analytics and artificial intelligence based analytics in hospitality and tourism management research. *International Journal of Contemporary Hospitality Management*. ISSN 0959-6119. https://doi.org/10.1108/IJCHM-08-2022-1006

Melian-Alzola, L., Fernandez-Monroy, M., & Hidalgo-Penate, M. (2020). Hotels in contexts of uncertainty: Measuring organizational resilience. *Tourism Management Perspectives, 36*(2020), 100747. https://doi.org/10.1016/j.tmp.2020.100747

Molina-Castillo, F.-J., Meroño-Cerdán, A.-L., Lopez-Nicolas, C., & Fernandez-Espinar, L. (2023). Innovation and technology in hospitality sector: Outcome and performance. *Business, 3*(1), 198–219. https://doi.org/10.3390/businesses3010014

Molina-Castillo, F. J., Rodríguez, R., López-Nicolas, C., & Bouwman, H. (2022). The role of ERP in business model innovation: Impetus or impediment. *Digital Business, 2*(2), 100024. https://doi.org/10.1016/j.digbus.2022.100024

Moriarty, J., Jones, R., Rowley, J., & Kupiec-Teahan, B. (2008). Marketing in small hotels: A qualitative study. *Marketing Intelligence & Planning, 26*(3), 293–315. https://doi.org/10.1108/02634500810871348

Motta, V. (2017). Are SMEs in the hospitality industry less likely to experience credit constraint than other industries in the service sector? Evidence from Latin America. *Tourism Economics, 23*(7), 1398–1418. https://doi.org/10.1177/1354816617692477

Motta, V., & Sharma, A. (2020). Lending technologies and access to finance for SMEs in the hospitality industry. *International Journal of Hospitality Management, 86*(2020), 1–9. https://doi.org/10.1016/j.ijhm.2019.102371

Mukherjee, S., Baral, M. M., Venkataiah, C., Pal, S. K., & Nagariya, R. (2021). Service robots are an option for contactless services due to the COVID-19 pandemic in the hotels. *Decision, 48*(4), 445–460. https://doi.org/10.1007/s40622-021-00300-

Nadkarni, S., Kriechbaumer, F., Rothenberger, M., & Christodoulidou, N. (2020). The path to the hotel of things: Internet of Things and Big Data converging in hospitality. *Journal of Hospitality and Tourism Technology, 11*(1), 93–107. https://doi.org/10.1108/JHTT-12-2018-0120

Nam, K., Dutt, C., Chathoth, P., Daghfous, A., & Khan, M. (2021). The adoption of artificial intelligence and robotics in the hotel industry: Prospects and challenges. *Electronic Markets, 31*, 553–574. https://doi.org/10.1007/s12525-020-00442-3

Nambisan, S. (2017). Digital entrepreneurship: Toward a digital technology perspective of entrepreneurship. *Entrepreneurship Theory and Practice, 41*(6), 1029–1055. https://doi.org/10.1111/etap.12254

Ntounis, N., Parker, C., Skinner, H., Steadman, C., & Warnaby, G. (2022). Tourism and hospitality industry resilience during the Covid-19 pandemic: Evidence from England. *Current Issues in Tourism, 25*(1), 46–59. https://doi.org/10.1080/13683500.2021.1883556

Ortiz-de-Mandojana, N., & Bansal, P. (2016). The long-term benefits of organizational resilience through sustainable business practices. *Strategic Management Journal, 37*(8), 1615–1631. https://doi.org/10.1002/smj.2410

Popesku, J., & Zaitanov, S. (2019). Current applications of artificial intelligence in tourism and hospitality. In *Sinteza 2019 – International scientific conference on information technology and data related research* (pp. 84–90). Singidunum University.

Prayag, G., Jiang, Y., Chowdhury, M., Hossain, M. I., & Akter, N. (2023). Building dynamic capabilities and organizational resilience in tourism firms during COVID-19: A staged approach. *Journal of Travel Research*. https://doi.org/10.1177/00472875231164976

Presenza, A., Abbate, T., Cesaroni, F., & Appio, F. P. (2019). Enacting social crowdfunding business ecosystems: The case of the platform Meridonare. *Technological Forecasting and Social Change, 143*, 190–201. https://doi.org/10.1016/j.techfore.2019.03.001

Priyono, A., Moin, A., & Putri, V. N. A. O. (2020). Identifying digital transformation paths in the business model of SMEs during the COVID-19 pandemic. *Journal of Open Innovation: Technology, Market, and Complexity, 6*, 104. https://doi.org/10.3390/joitmc6040104

Rahimi, R. (2017). Customer relationship management (people, process and technology) and organizational culture in hotels: Which traits matter? *International Journal of Contemporary Hospitality Management, 29*(5), 1380–1402.

Rahimi, R., Akgunduz, Y., & Bilgihan, A. (2023). Impact of COVID-19 on mental health and career anxiety of hospitality and tourism students in the UK. *Journal of Hospitality and Tourism Insights, 6*(2), 892–911.

Rahimi, R., & Gunlu, E. (2016). Implementing customer relationship management (CRM) in hotel industry from organizational culture perspective: Case of a chain hotel in the UK. *International Journal of Contemporary Hospitality Management, 28*(1), 89–112.

Rahimi, R., Köseoglu, M. A., Ersoy, A. B., & Okumus, F. (2017). Customer relationship management research in tourism and hospitality: A state-of-the-art. *Tourism Review, 72*(2), 209–220.

Rahimi, R., & Kozak, M. (2017). Impact of customer relationship management on customer satisfaction: The case of a budget hotel chain. *Journal of Travel & Tourism Marketing, 34*(1), 40–51.

Reficco, E., Layrisse, F., & Barrios, A. (2021). From donation-based NPO to social enterprise: A journey of transformation through business-model innovation. *Journal of Business Research, 125*, 720–732. https://doi.org/10.1016/j.jbusres.2020.01.031

Reinmoeller, P., & Van Baardwijk, N. (2005). The link between diversity and resilience. *MIT Sloan Management Review*. https://sloanreview.mit.edu/article/the-link-between-diversity-and-resilience/

Rezaei, S., Shahijan, M. K., Valaei, N., Rahimi, R., & Ismail, W. K. W. (2018). Experienced international business traveller's behavior in Iran: A partial least squares path modelling analysis. *Tourism and Hospitality Research, 18*(2), 163–190.

Sarmah, B., Kamboj, S., & Rahman, Z. (2017). Co-creation in hotel service innovation using smart phone apps: An empirical study. *International Journal of Contemporary Hospitality Management, 29*(10), 2647–2667. https://doi.org/10.1108/IJCHM-12-2015-0681

Sharma, U., & Gupta, D. (2021, July). Analyzing the applications of internet of things in hotel industry. *Journal of Physics: Conference Series, 1969*(1), 012041. IOP Publishing.

Shin, H., Jeong, M., & Cho, M. (2021). The impact of smart tourism technology and domestic travelers' technology readiness on their satisfaction and behavioral intention: A cross-country comparison. *International Journal of Tourism Research, 23*(5), 726–742. https://doi.org/10.1002/jtr.2437

Sigala, M., Rahimi, R., & Thelwall, M. (2019). Big data and innovation in tourism, travel, and hospitality. *Culinary Science & Hospitality Research, 23*(6), 1–11.

Somers, S. (2009). Measuring resilience potential: An adaptive strategy for organizational crisis planning. *Journal of Contingencies & Crisis Management, 17*(1), 12–23. https://doi.org/10.1111/j.1468-5973.2009.00558.x

Statista. (2022). Market size of the smart hospitality sector worldwide in 2020, with a forecast for 2027, May 2022 Available at: https://www.statista.com/statistics/1252565/global-smart-hospitality-market-size/

Stylos, N., Fotiadis, A. K., Shin, D., & Huan, T. C. (2021). Beyond smart systems adoption: Enabling diffusion and assimilation of smartness in hospitality. *International Journal of Hospitality Management, 98*(2021), 103042. https://doi.org/10.1016/j.ijhm.2021.103042

Stylos, N., & Vassiliadis, C. A. (2023). Gamification design: Toward developing image perception scales for generation Z consumers. *International Journal of Contemporary Hospitality Management*. https://doi.org/10.1108/IJCHM-12-2022-1616

Stylos, N., & Zwiegelaar, J. (2019). *Big data as a game changer: How does it shape business intelligence within a tourism and hospitality industry context?* (pp. 163–181). Springer.

Stylos, N., Zwiegelaar, J., & Buhalis, D. (2021). Big data empowered agility for dynamic, volatile, and time-sensitive service industries: The case of tourism sector. *International Journal of Contemporary Hospitality Management, 33*(3), 1015–1036.

Teece, D. J. (2010). Business models, business strategy and innovation. *Long Range Planning, 43*(2–3), 172–194. https://doi.org/10.1016/j.lrp.2009.07.003

The EU Data Protection Reform and Big Data. (2016). https://op.europa.eu/en/publication-detail/-/publication/51fc3ba6-e601-11e7-9749-01aa75ed71a1

Urbaniec, M., & Żur, A. (2021). Business model innovation in corporate entrepreneurship: Exploratory insights from corporate accelerators. *The International Entrepreneurship and Management Journal, 17*, 865–888. https://doi.org/10.1007/s11365-020-00646-1

Vassiliadis, C. A., & Stylos, N. (2022). Data mining. In D. Buhalis (Ed.), *Encyclopedia of tourism management and marketing* (pp. 790–792). Edward Elgar Publishing.

Vatankhah, S., Darvishmotevali, M., Rahimi, R., Jamali, S. M., & Ebrahim, N. A. (2023). Assessing the application of multi-criteria decision making techniques in hospitality and tourism research: A bibliometric study. *International Journal of Contemporary Hospitality Management, 35*(7), 2590–2623.

Verreynne, M. L., Williams, A. M., Ritchie, B. W., Gronum, S., & Betts, K. S. (2019). Innovation diversity and uncertainty in small and medium sized tourism firms. *Tourism Management, 72*, 257–269. https://doi.org/10.1016/j.tourman.2018.11.019

Vogus, T. J., & Sutcliffe, K. M. (2007, October). Organizational resilience: Towards a theory and research agenda. In *2007 IEEE international conference on systems, man and cybernetics* (pp. 3418–3422). IEEE.

Weerawardena, J., Mort, G. S., Salunke, S., & Haigh, N. (2020). Editorial and research agenda: JBR special issue on business model innovation in social purpose organizations. *Journal of Business Research, 125*, 592–596. https://doi.org/10.1016/j.jbusres.2020.07.032

Yallop, A., & Seraphin, H. (2020). Big data and analytics in tourism and hospitality: Opportunities and risks. *Journal of Tourism Futures, 6*(3), 257–262. https://doi.org/10.1108/JTF-10-2019-0108

Yang, H., Song, H., Cheung, C., & Guan, J. (2021). How to enhance hotel guests' acceptance and experience of smart hotel technology: An examination of visiting intentions. *International Journal of Hospitality Management, 97*. https://doi.org/10.1016/j.ijhm.2021.103000

Zhang, H., Zang, Z., Zhu, H., Uddin, M. I., & Amin, M. A. (2022). Big data-assisted social media analytics for business model for business decision making system competitive analysis. *Information Processing and Management, 59*(1). https://doi.org/10.1016/j.ipm.2021.102762

Zott, C., Amit, R., & Massa, L. (2011). The business model: Recent developments and future research. *Journal of Management, 37*(4), 1019–1042. https://doi.org/10.1177/0149206311406265

5

Hybrid Servicescapes in Tourism and Hospitality

Ellis Urquhart

Introduction

Bitner (1992, p. 65) defined the servicescape as "*all of the objective physical factors that can be controlled by the firm to enhance (or constrain) employee and customer actions.*" These include a series of environmental dimensions, namely ambient conditions, space/function, and signs, symbols, and artefacts. In combination, the elements of the servicescape can generate internal responses for customers and employees and influence subsequent behaviours. In the contemporary tourism and hospitality landscape, digital resources and ICTs (such as automated touchpoints, smartphone applications, augmented reality, and robotics) have increasingly been explored within the customer experience, suggesting a need to reframe our understanding of the servicescape.

The reconfiguration of the servicescape is particularly relevant in the post-Covid-19 landscape. Serrano-Baena et al. (2023) highlighted the

E. Urquhart (✉)
Edinburgh Napier University, Edinburgh, UK
e-mail: e.urquhart@napier.ac.uk

N. Stylos et al. (eds.), *Contemporary Marketing Management for Tourism and Hospitality*, https://doi.org/10.1007/978-3-031-65049-9_5

alteration of tourism and hospitality environments to incorporate government restrictions and the addition of new features such as face masks, sanitation stations, and increased distancing as a key driver that changed consumer perceptions towards service design. Similarly, due to the rapid digitisation and adoption employed across the service sector during Covid-19 (Utkarsh & Sigala, 2021), digital resources now represent an inherent feature in the contemporary (and future) tourism and hospitality experiences. The dominance of self-service capabilities, smartphone technology, and augmented and virtual reality have extended the traditional servicescape with digital resources stimulated by a re-evaluation of technology adoption in the post-pandemic landscape (Iskender et al., 2022). Furthermore, the growing investment in robotics, automation, smart surfaces, and interactive touchpoints provides further opportunities for the hybridisation of the servicescape. Similarly, the presence of digital resources within the service environment affords enhanced customer autonomy and scope for customers to actively co-create their own experiences (Hussain, 2021) and the subsequent value derived from them (Massari et al., 2022). Yet, little research has actively considered the influence of the hybrid servicescape on the co-creation of tourism and hospitality experiences.

The purpose of this chapter is to explore the increasing hybridisation of the servicescape in tourism and hospitality environments. Drawing on the seminal contributions of Bitner (1992) and further adaptions of the servicescape concept, this chapter considers the nexus between physical, ambient, and digital resources within the experiential space. Furthermore, these hybrid servicescapes are examined through a co-creative lens to debate the implications for the customer experience. Namely, the work investigates how hybrid spaces can facilitate active dialogue, engagement, personalisation, and customer control within tourism and hospitality businesses.

Revisiting the Servicescape

Despite being over 30 years since the publication of Bitner's (1992) servicescape model, the lasting interest in the role of physical evidence in service organisations remains strong. At the heart of the concept is the link between physical elements of the service environment and the internal responses that they create for employees and customers. Thereafter, the influence on customer and employee behaviours as a result of the servicescape is explored. Beyond having an impact on customers and employees, Bagchi et al. (2021) highlight the importance of the servicescape and its management to sustain tourism and hospitality businesses within a highly competitive market. Conversely, the need to adapt and innovate individual servicescapes based on changing consumer preferences and expectations is a lasting challenge for businesses.

Environmental Dimensions

Central to Bitner's (1992) servicescape model are environmental dimensions: ambient conditions; spatial layout and functionality; and signs, symbols, and artefacts. At the ambient level, elements of design which activate the senses can be seen as a critical part of the customer and employee experience. These include the air quality and temperature which can create a hot or cold environment. Additionally, the intensity and colour of lighting can activate visual sensations and can evoke certain responses. For example, Bilgili et al. (2020) identified that the combination of lighting colours within a hospitality setting can affect how customers perceive waiting times. And how adapting the colour palette and intensity can mitigate negative feelings about the wait time. Sound and background noise also align with this dimension and act as a significant feature of tourism and hospitality. Exposure to music, ambient sounds, or soundscapes can shape customer responses within a space. As noted by Lu et al. (2022), sound can be manipulated to create a particular image or ambience within spaces. The range of sound, volume, pitch, and frequency can be adapted to evoke calm, stimulate engagement, or inform (e.g. announcements).

The second dimension refers to spatial layout and functionality. The layout of the physical space, including machinery, equipment, and furnishings can create immediate impressions for customers and employees (Bitner, 1992). For example, building designs in angular shapes (e.g. square with straight lines and sharp edges) have been found to evoke perceptions of harshness, importance, and seriousness in the minds of customers. Conversely, buildings/features in circular shapes can be perceived as more social, inclusive, and gentle (Liu et al., 2018). In tourism and hospitality, even small elements of layout can have significance for customers and staff. For example, as noted by Ryu and Park (2019), the availability, comfort, and positioning of seating areas within airport spaces can create high levels of satisfaction for travellers, while conversely, a lack of seating or the proximity of seats can dissatisfy customers with increased dwell times. This also came into sharp focus during the Covid-19 pandemic where distancing regulations altered the layout within public-facing spaces such as hospitality (Lestari et al., 2021). At a functional level, the layout of the space can support or inhibit efficiency, enjoyment, and customer/staff movement which equally contribute to wider responses and behaviours. From a co-creative perspective, Kim and Baker (2022) found that partitions (e.g. barriers, dividers) and signage help to establish customer norms and evoke trust in the competence of the organisation. Both features are central to fostering co-creative behaviour where customers need to have guidance on how to navigate and use the service space effectively.

Finally, signs, symbols, and artefacts can have a powerful impact on the first impressions of customers and employees (Bitner, 1992). Often used to promote or limit behaviours (e.g. emergency exit signs, no smoking labels), signs and symbols provide a frame of reference for customers within a physical space. Furthermore, signs and symbols are particularly relevant for supporting customer information, wayfinding, and behaviours. This is complicated by the unique knowledge, abilities, and preferences of individual customers. For example, Wan (2022) considered the accessibility of tourist signage and the importance of universal design principles to support equal access for diverse tourist backgrounds and capabilities. Tangible artefacts such as business cards, letterheads, and stationery can also reinforce brand imaging and create consistency in design within a physical service environment.

Internal Responses

As per Bitner's (1992) assertion, the combination of physical evidence and its design/use in service environments can therefore generate internal responses in both customers and employees. From an environmental psychology perspective, these can fall into cognitive, emotional, and physiological responses. Firstly, cognitive responses can be generated in response to the physical or ambient cues within the space. These can then influence beliefs about the service (and its offering) in the minds of customers and employees. For example, elements of design within the physical space can influence beliefs of trustworthiness about an organisation (Lai & Chong, 2020). Small aspects, such as physicians wearing a stethoscope at all times, provide a visual cue and can stimulate the belief that the organisation is prepared and well-equipped for urgent care. Similarly, the presence of legal texts within law offices and solicitor consultation rooms may stimulate beliefs around competence and knowledge sought out by clients. From an employee perspective, the physical space can inspire beliefs about the nature of individual roles and teams within an organisation (Tran Xuan et al., 2022). For example, an open-plan office design can evoke beliefs of a cohesive team-based community, whereas individual closed offices can stimulate beliefs of autonomy, authority, and power.

The environment can also elicit emotional responses from customers and employees. As Bitner (1992) suggested, not only does the composition of the space and environmental factors influence how people think (cognition), but it can also influence how they feel. Namely, perceptions of the environment can generate feelings of pleasure (of displeasure) and arousal (excitement and stimulation). For customers, pleasing aesthetics, functionality, and ambience can lead to a pleasure response which, in turn, can inspire a willingness to stay within the space for longer periods. This can be illustrated in Fig. 5.1 with 'The Jewel' attraction within Singapore Changi International Airport. As shown, this space demonstrates a design concept that integrates functionality with aesthetic appeal in a space that typically has an increased dwell time for travellers. Such an approach aligns with Antwi et al. (2022) assertion that future airports

Fig. 5.1 'The Jewel' at Singapore Changi Airport (Source: Wiki Commons, Public Domain, 2023)

need to be staged with a mixture of sensory and culturally sensitive features to enhance passenger satisfaction. While degrees of arousal and sensory stimulation can have positive effects on customers, overstimulation can discourage customers from remaining in the space (Lin & Mattila, 2010; Lin & Worthley, 2012). Examples of overstimulation would include excessive or stark lighting, increased noise or background music, or a chaotic space that lacks functionality. For employees, similar feelings of pleasure/arousal can be influenced by the working environment. A well-designed space that is not only functional to complete duties but also aesthetic with attention to employee comfort can lead to feelings of pleasure. Equally, poorly designed staff spaces that do not support employee well-being and productivity can produce negative emotions which influence behaviour.

Finally, there is evidence to suggest that the physical and ambient environment can generate physiological responses in customers and employees (Bitner, 1992). As with the previous elements, the environment can stimulate both positive and negative responses. A well-designed space and suitable ambient conditions, such as adjustable temperature and effective ventilation, can create a comfortable space for individuals to engage with, enhancing a feeling of well-being (Lee & Chuang, 2022). Equally, poorly designed spaces can create physiological challenges. For example, excessive heat or low temperature can be discomforting for customers causing them to shiver or perspire. High-pitch frequencies or high-treble background music can be damaging to ear health over extended periods (Novak et al., 2010). Bright lighting or strobe effects can be not only displeasing but also a potential trigger for customers with epilepsy and light sensitivity. Finally, the design of building features can create problems for customer and employee access. This could be found in narrow staircases, polished floors, or poorly designed walkways. From an employee perspective, there may be a heightened impact of the physical environment on physiology. Due to the length of exposure being higher than customers, the physical surroundings may support or potentially damage the health and well-being of staff (Kazlauskaite et al., 2022). Examples include the ergonomic design of workspaces and units, favourable lighting, ambient conditions, and physical features that prevent strain, such as dimmable PC monitors and appropriate handheld devices.

Behaviour

The output of the servicescape model is the behaviours exhibited by customers and employees that have been influenced by environmental conditions. Mehrabian and Russell (1976) split these behaviours into two categories: approach and avoidance behaviour. For customers, a positive response to the physical environment can influence a range of approach behaviours. Well-designed servicescapes can encourage customers to engage with the service initially in addition to extending their exposure to the space. This may take the form of extended dwell time in a favourable space. Equally, with increased stay there is potential for increased spending on-site and a greater likelihood of the customer returning the service with loyalty (Han & Ryu, 2009). Interestingly, Lin et al. (2019) found that millennial visitors valued the social density of a space (i.e. the presence of other visitors and their interactions) to have a more important influence on revisit intention, indicating differing perceptions of the servicescape for different generational cohorts. This reiterates the need for contemporary service managers to integrate opportunities for peer socialisation within the physical design of service spaces (Zhang et al., 2023). Conversely, poorly designed service environments can generate customer avoidance behaviours. This may discourage customers from entering the space or limit the extent they wish to stay and explore. As a result of a reduced dwell time, customers are less likely to spend extra on additional services and are less likely to return to the service again in the future.

For employees, positive responses to the servicescape in which they work can create approach behaviours such as stronger affiliation with the organisation and brand. Equally, like customers, employees who are satisfied with the physical surroundings are more likely to demonstrate greater commitment to their role and duties in addition to remaining with the organisation long term. In contrast, poorly designed employee servicescapes can lead to demotivation and reduced affiliation with the organisation and their role. From a functional perspective, the physical design may inhibit features of the employee role and can lead to employee turnover. This is particularly relevant for service managers as there is typically lower attention to detail in the aesthetics and functionality in employee-only (back-of-house) spaces (Kirillova et al., 2020).

Extending the Servicescape Model

Rosenbaum and Massiah (2011) expanded the traditional servicescape model to capture additional stimuli and dimensions. The authors argued that beyond the physical assets, social, socially symbolic, and natural dimensions contributed to the perceived servicescape. At a social level, it has been argued that the social interactions between stakeholders within the service experience can have a profound impact on the customer and employee experience (Line & Hanks, 2019). Social interactions, customer/employee image, and the nature of the crowd within a space can influence cognitive, affective, and approach/avoidance reactions (Pizam & Tasci, 2019; Tasci & Pizam, 2020). The socially symbolic dimension involves the presentation of signs, symbols, and artefacts that carry socio-collective meaning and significance to different social groups (Rosenbaum & Massiah, 2011). As discussed by Verhoeven et al. (2009), individual customers can associate subjective symbolic associations with various elements of service designs. For example, neon lighting in a restaurant setting can trigger symbolic perceptions of the fast-food industry, whereas candlelight can be associated with the luxury à la carte sector. Finally, natural dimensions include the design of servicescapes that utilise natural resources such as animal and/or plant time and the associated customer responses to these. As argued by Rosenbaum and Massiah (2011), the presence and use of natural resources can evoke feelings of wellness, restoration, and calm which may form part of the wider service experience. For example, the presence of living plants in hospitality settings has been found to enhance perceived experiential value through aesthetic appeal (Apaolaza et al., 2020). Similarly, Jang (2021) argued that a 'green servicescape' with an environmentally conscious design can enhance customer attachment and loyalty to a service brand.

Integrating the Digital Resource

There have been theoretical contributions aiming to integrate digital technology into the servicescape perspective. Harris and Goode (2010) explored the growth of the 'online servicescape' due to mass access to the Internet and online marketplaces that facilitate contact between customers and providers. Similarly, Tankovic and Benazic (2018) presented the 'e-servicescape' as a significant point of transition from physical experiences into the online space. Equally, Allmér (2020) draws distinctions between the physical servicescapes and digital ones due to the differences in sensory interactions that are available to customers. In the retail context, Roy et al. (2019) introduced the concept of the 'smart servicescape' that identified key dimensions related to the role of smart devices in assisting consumers in their shopping goals and the connection with brand equity. Similarly, Amitrano et al. (2021) considered the 'augmented servicescape' in a heritage setting and the potential for augmented and virtual technology to extend the boundaries of the physical museum. Research has also captured the mobile dimension. Lee (2018) explored the 'm-servicescape' and highlighted the value of well-designed mobile applications in the hotel sector for customer engagement and developing brand loyalty. Finally, Yin et al. (2023) recently defined the 'intelligent servicescape' as the integration of artificial intelligence (AI) technology, such as sensors and devices, into physical spaces. However, much of the existing research creates a firm distinction between physical and digital (i.e. online) servicescapes in addition to being explored in other contexts which differ from the customer motivations and experiences within tourism and hospitality settings. This chapter advocates for the hybridisation of the physical and digital within the on-site tourism or hospitality space. This experiential convergence merges digital resources with the tangible landscape to add additional sensory layers to the customer experience (Urquhart, 2019). In extending these studies, Fig. 5.2 provides an overview of contemporary servicescape components with the inclusion of digital resources onsite.

Within the *physical category*, traditional elements of space and function remain critical within the tourism and hospitality landscape. While there has undoubtedly been growth in purely online services, physical and tangible aspects remain a significant element of the customer experience. Han (2013) argued that the arrangement of amenities, layout, and functionality dimensions have a significant impact on customer comfort, efficiency, and satisfaction in extended-duration activities (such as the in-flight experience). Similarly, signs, symbols, and artefacts still feature heavily in tourism and hospitality settings. In the visitor attraction sector, for instance, directional signage and interpretative labelling represent a key part of the visitors' route and their understanding of the attraction story. These physical cues can go beyond information provision into supporting the visitor's journey through leisure activities. An example would include the provision of visitor trails in natural or cultural spaces. Not only does this provide a linear sequence for visitors to follow but also

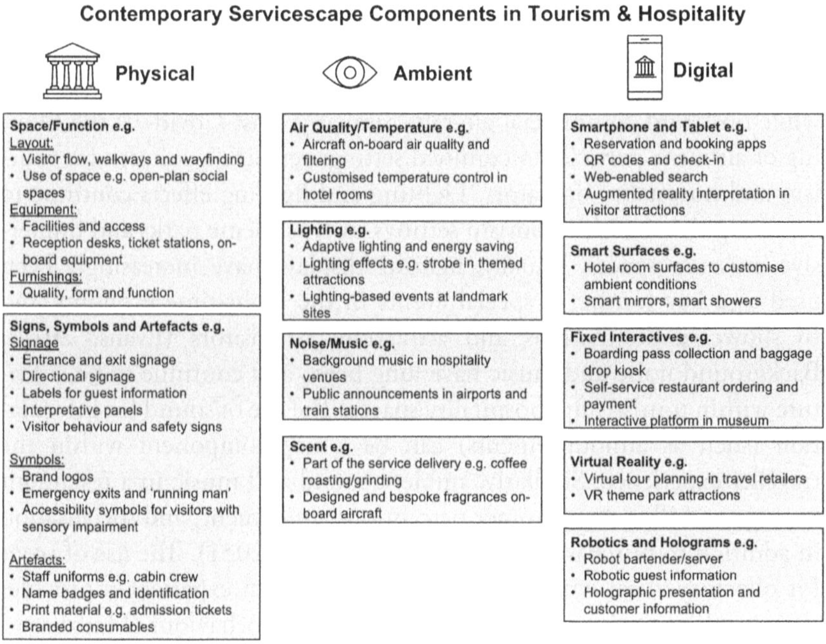

Fig. 5.2 Contemporary servicescape components in tourism and hospitality (Source: Author)

contextualises information into a manageable story for visitors to engage with (MacLeod, 2017). Physical signage and information can also influence visitor behaviour, and attitude, and reinforce key messages for the site such as nature conservation within National Parks (Juma et al., 2020). Equally tangible artefacts, such as staff uniforms, remain a key feature in hospitality venues and the transport sector. As discussed by Choy and Shih (2022), the presence of uniforms not only establishes role behaviour for employees but has impact on the perception of quality and trust for customers in hospitality. Interestingly, when exploring this in a generational context, Goh and Lee (2018) found that among Gen Z, the presence of uniforms and professional appearance within hospitality spaces evoked a sense of glamour. As such, the future design of service uniforms will need to consider the style-conscious motivations of younger and emerging generational cohorts.

Ambient components also feature and can be tailored within a variety of tourism and hospitality settings. For example, air quality, temperature, and ventilation are key elements of passenger comfort and well-being while on-board commercial aircraft. Similarly, post-Covid-19 the filtering of airborne particles in confined settings remains a concern for tourism and hospitality operators. Lighting and lighting effects continue to be manipulated within tourism settings such as theme parks and immersive museum spaces. Equally, lighting displays have increasingly been used as a key feature of special events in the night-time economy and to showcase destinations and attractions to visitors (Evans, 2012). Background noise and music have long been, and continue to be, a feature within tourism and hospitality spaces. The use of sound for information (such as announcements) can be a key component within the traveller experience. Similarly, ambient sound and music in a restaurant setting can influence customer perceptions, enjoyment, and socialisation in addition to purchase behaviour (Tarlao et al., 2021). The use of scent (or olfactory cues) within hospitality and tourism experiences provides opportunities to shape visitor memorability and behaviour (Madzharov, 2022). This may take the form of associated scents in the environment as a result of the service delivery. Examples include the grinding of coffee beans within Starbucks store units or food aromas within restaurant settings. Similarly, scent can be purpose designed and used as part of the servicescape. Examples such as the designed fragrance, Stefan Floridian Waters, on Singapore Airlines which has increasingly become synonymous with the brand through its use on-board and in passenger amenities (Errajaa et al., 2021). The construction and active design of the ambient conditions continue to be important for service managers and their target markets. For example, Peng et al. (2022) highlighted the important role that ambience has in the service perceptions and satisfaction of Millennial tourists in China.

Increasingly, the division between the physical and *digital* resources within tourism and hospitality spaces is becoming more blurred. Smartphone and tablet-enabled interfaces within tourism and hospitality settings have become the norm. While these offer numerous opportunities for customers to engage with the organisation pre- and post-consumption; smartphone and tablet technology can routinely be observed within the physical service space. As discussed by Lee (2018) smartphone applications have replaced several tangible elements within hospitality, such as removing the need for room keys and physical loyalty cards. Similarly, standalone applications within hospitality providers allow for booking, amendments, guest requests, and real-time communication. Similarly, quick response (QR) codes that can be read by smartphone and tablet provide opportunities to link to content, visitor information, and booking options and are an effective tool for capturing on-site feedback from visitors within destinations (Vuksanovic et al., 2021). During Covid-19, these also became key features in transport spaces through the introduction of QR health certification and vaccination reporting (Zheng et al., 2022). Augmented reality (AR) has also altered the way customers can engage with tourism and hospitality spaces. Whilst retaining a physical presence in the space but augmenting content through smart devices, AR can blend the real world with digital layers (Yung & Khoo-Lattimore, 2019). For example, within the heritage sector, AR provides an opportunity of overlaying information about historical assets whilst not overcrowding or disrupting the space (Yin et al., 2021) and can facilitate engagement between visitors and exhibitions at a safe distance (Urquhart, 2022).

Additional digital tools that are becoming more common in tourism and hospitality settings include smart surfaces. These surfaces utilise physical elements that would typically feature in hospitality spaces but use them as a digital interface. An emerging example is the 'Smart Mirror', where hotel guests can not only use this as a mirror but also interact with it through both touch and voice activation to access applications, hotel functions, and place requests (Fox, 2018). This conversion of physical amenities to multi-use digital tools may be a significant trend in the future of hospitality servicescapes and other sectors (Ristova & Dimitrov, 2019). Fixed interactive tools are well-established in a range of tourism and hospitality spaces. Many of these take the form of self-service technologies to provide greater autonomy, convenience, and efficiency to customers (Kelly & Lawlor, 2021). Nevertheless, fixed interactives feature prominently within the physical servicescape. These may act as points of transaction, such as ordering and payment terminals within fast-food retail spaces, as a point of collection such as boarding pass printing terminals in airport departure halls, or as an element of the experience design such as interactive units within museums and science centres. More advanced still is the presence of virtual reality (VR) experiences in tourism and hospitality. While the nature of VR does not require a physical presence (i.e. individuals can take part in VR activities without entering a physical space), there is evidence of VR tools being used within physical leisure spaces as part of the customer offering. Lee and Kim (2021) highlight that VR can be used to immerse individuals in digitally reconstructed spaces but equally, there can be VR experiences alongside the physical. Examples from the theme park sector include the 'Space Fantasy' VR roller coaster at Universal Studios, Japan, or the 'Arkham Asylum: Shock Therapy' VR coaster at Warner Bros. Movie World, Australia, blending physical surroundings with a virtual experience. Similarly, VR can be used to inform and illustrate intangible activities to aid the marketing function (Wei, 2019). An example can be seen in the Heritage Collection Travel retail space in York, UK, where the travel agency features a high-street presence with a virtual reality cinema room on-site. This space allows clients to virtually visit destinations as part of the consultation to help visualise the experience and inform their booking options (Bush, 2022).

Finally, the implementation of robotics and holograms within tourism and hospitality settings continues to alter the onsite servicescape. While the introduction of robots in tourism and hospitality lagged other industries due to the limitations in how they could react and respond to individual customer needs, there has been significant growth in this area (Ivanov et al., 2019). This has, however, posed challenges for the servicescape. As suggested by Manthiou et al. (2021), tourism and hospitality managers need to consider the implications of robotic adoption within existing tangible servicescapes. As public-facing spaces, the integration of robotic assets can create challenges on a physical level, such as difficulties in manoeuvring robotic components, overcrowding, and potential collision with customers and employees. Furthermore, from an environmental perspective, the design and functionality of robotic assets within the servicescape will have a significant impact on the customer experience. As with wider digital applications, customer acceptance, trust, and use are linked to the functionality and value that can be provided to customers (Lu et al., 2020). Merdin-Uygur and Ozturkcan (2023) found that although robotic applications were more positively received by Gen Z consumers, the changeable attitudes of this cohort posed challenges for service providers as to the long-term design planning of service spaces. Nevertheless, there are several examples of robotic applications within popular tourism and hospitality environments. For instance, the temporary deployment of UniPi robot Centro Piaggio in Pisa (see Fig. 5.3). Equally, robotics can be found within spaces as a feature of service delivery to replace employee actions. An example can be found in the 'Bionic Bar' onboard Royal Caribbean Cruises featuring two robotic bartenders that can create bespoke cocktails based on passenger preference. Similarly, holographic technology is becoming more visible within the servicescape. These often take the form of presentations, information points, or other

Fig. 5.3 UniPi robot Centro Piaggio: humanoid robot application (Source: Wiki Commons, 2015)

virtual content that can be integrated into the physical space (Hammady et al., 2021). Examples from the sector include holographic gun control presentations at Atlanta International Airport to support the security function or the holographic passenger assistance terminal in Dubai Airport lounges to provide traveller information.

The Hybrid Servicescape as a Co-creative Space

The concept of co-creation has emerged from service-dominant logic where value is generated through 'use' (as opposed to exchange) and an integration of various tangible (operand) and intangible (operant) resources (Vargo & Lusch, 2004). Co-created experiences emerge as a result of customers activating their resources and acting as a generator of their own value within the experience (Sfandla & Björk, 2013). As such, service providers act as a facilitator for experience co-creation and can design the service space, initiate the interactions, and provide co-creative platforms, interfaces, and artefacts to support the process (Ramaswamy & Ozcan, 2018). Considering the role of the servicescape, Prayag et al. (2020) highlights the importance of the physical environment to support co-creative experiences, noting that carefully designed spaces can be stimulating to the senses and establish interactive opportunities. As discussed by Ramaswamy and Ozcan (2020), digital platforms and interfaces act as a mediating force between actors (customers and service providers) to support meaningful interactions within experiences. Similarly, Roy et al. (2019) argue that smart consumers of today highly value the personalisation benefits of technological mediation in service spaces. In drawing these concepts together, Fig. 5.4 connects the hybrid servicescape with four key co-creative opportunities. This chapter suggests that the combination of physical, ambient, and digital resources that customers engage with can provide a series of co-creative opportunities that have value-added applications in tourism and hospitality spaces. The following sections explore each of these opportunities in turn and consider how these may be applied at a management level.

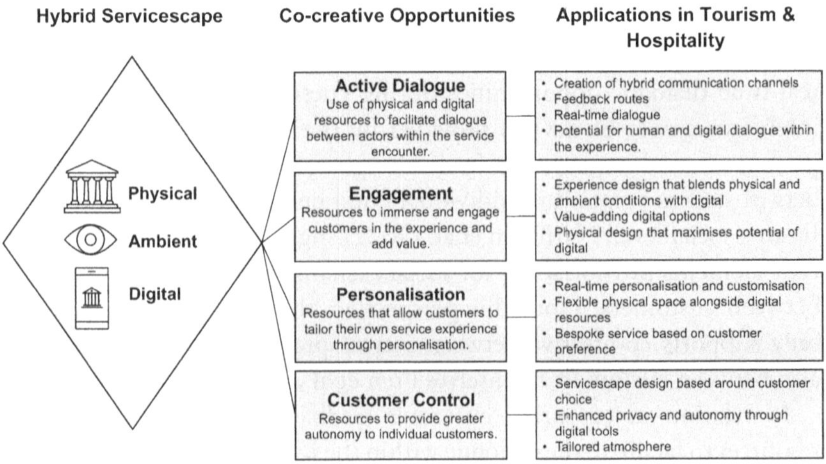

Fig. 5.4 Activating the hybrid servicescape to facilitate co-creative opportunities in tourism and hospitality (Source: Author)

Active Dialogue

As discussed by Vargo and Lusch (2004), the flow of active dialogue between actors in the service encounter is a fundamental premise in the co-creation of experiences. Within a co-creative ethos, this can take the form of customer-to-business dialogue as well as opportunities for customer-to-customer dialogue within social experiences (Rihova et al., 2018). Applying this to the hybrid servicescape involves the use of various physical and digital resources to facilitate a meaningful dialogue between actors in the service encounter. Grönroos and Voima (2013) referred to this as the 'joint sphere', the space in which value is created through dialogical interaction. In this context, the service provider acts as a value facilitator and designs the space and platforms to encourage dialogue and interaction between actors.

In tourism and hospitality, various practices utilise both physical and digital resources to facilitate dialogue between customers and employees. Many of these are used to collect and respond to customer preferences. In the hotel environment, while booking preferences can be identified pre-arrival, there are examples of digital resources that can be used during the experience to adapt requests and preferences onsite. Instant guest messaging services and digital concierge platforms (such as GuestTouch) provide real-time dialogue opportunities during guest stays. Similarly, with the interconnectivity of devices provided by the Internet of Things (IoT), preferences and requests can feed through other digital tools and members of staff to respond to individual guest enquiries (Chapman, 2017). From a social level, Nyrhinen et al. (2022) suggested that digital servicescape elements provide scope for social exchanges and virtual interactions between customers. From this perspective, the hybrid servicescape not only supports co-creative between service providers and customers but also between customers themselves (Yen et al., 2020).

In relation to capturing customer feedback, there are a variety of resources to facilitate the dialogue within the hybrid servicescape. Physical evidence such as comment cards and feedback forms provide cues to hotel guests to share views and metrics. Similarly, physical information points, receptions, and helpdesks within the space are key physical components to support face-to-face dialogue between customers and staff members. Interactive touchpoints such as physical feedback buttons, kiosks, and touchscreen evaluation tools can be seen in several tourism environments, such as airports and train stations. Such digital resources provide customer autonomy and the choice to share comments with anonymity. Similarly, handheld devices can be used to capture customer comments through feedback sections on organisational mobile applications. However, the potential of hybrid feedback provides further opportunities for meaningful dialogue. For example, customer feedback via tablet devices but instigated and facilitated by a member of staff in a joint interaction between the service provider and customer (Gremyr et al., 2022). This has the benefit of rapid data collection/analysis but also a personal quality to humanise the dialogue and, potentially, allow for real-time responses and service recovery.

Engagement

As discussed by Buhalis and Sinarta (2019) within the co-creative ethos, real-time dynamic engagement between actors within the service space can optimise value for stakeholders. In practice, tourism and hospitality providers can design their servicescape with various resources that can immerse and engage customers within the experience and, by association, add value. Equally, Oyner and Korelina (2016, p. 343) argued that for engaging experience to occur, service providers need to "*allow clients to choose and manage the crucial component of customer value* – [the] *customer experience.*" The hybrid servicescape is well-placed to support co-created engagement through a variety of physical, ambient, and digital resources within the service space.

Kim et al. (2020) argued that customer engagement can be strengthened by the careful design and staging of the physical service space. As such, dynamic experience design to facilitate engagement includes not only the skills and competencies of staff but also the tangible and intangible elements to provide engagement touchpoints. The Walt Disney Company have termed this 'guestology' where every element of the service production is designed in a customer-centric approach. The cast (staff), setting (space), and ambience/atmosphere are meticulously designed to support customer enjoyment and value (see Fig. 5.5). Within such an approach, a greater focus on the hybrid servicescape and value-adding digital options is a key priority for the tourism and hospitality sector.

Of key concern is the integration of the physical, ambient, and digital within existing spaces and the impact this has on the customer experience. As highlighted by Alyahya McLean (2022), customers form experiences at a sensory level, which can form attitudes and interpretations. From a management perspective, this creates a challenge for the hybrid design of service spaces. The combination of physical artefacts, ambient conditions, and digital tools needs to be designed in ways that activate multiple customer senses. Equally from a customer perspective, the design, aesthetic, and function of service spaces are key contributors to perceptions of satisfaction, value, and loyalty. An example of this is the rise of Instagramability within tourism spaces. Heo et al. (2023) suggest that, particularly for younger generations, the desire to take, upload, and share visually appealing images at tourist sites such as theme parks holds significant importance for contemporary visitor experiences and brand value. Therefore, the composition of the space needs to be designed

Fig. 5.5 'Sleeping Beauty Castle', Disneyland Paris (Source: Author)

to add relevant and value-adding touchpoints whilst guarding against sensory overload. This is particularly relevant given the rise of the 'metaverse' in tourism and hospitality. As suggested by Gursoy et al. (2022), there is potential for 3D avatar-based interactions to augment the physical experience. From a hybrid servicescape perspective, the potential for metaverse applications that extend the physical space is significant for customer engagement and immersion.

Personalisation

A central tenet of co-creative experiences is the potential for customers to personalise and tailor the experience that they are engaging with (Jain et al., 2021). Within the service sector, personalisation can occur at each stage of the service encounter. In the pre-experience stage, customers can personalise their booking options, requests, preferences, and details through online booking platforms, customer contact centres, and applications. During the in-situ experience, firms can combine various human, inanimate, and digital touchpoints to provide opportunities for customers to tailor the experience in real time. Finally, in the post-experience stage, customer relationship management systems provide opportunities for targeted feedback, personalised communications, and follow-up services (Chathoth et al., 2016). Essential to a co-creative experience is the ability for customers to engage in real-time personalisation during the service encounter. As discussed by Lei et al. (2019), this real-time personalisation extends into the control that customers have over their environment. From a hybrid servicescape perspective, flexibility in the physical, ambient, and digital space provides opportunities for customers to personalise their environment. For example, hotel rooms with where not only the physical assets can be moved into a bespoke arrangement, but ambient controls to tailor sensory conditions (e.g. room scent, lighting tone, and sound). Extending the notion of real-time personalisation, Haobin et al. (2021) advocated for hospitality service providers to integrate surprise features within their servicescapes for customers to engage in real-time mindful behaviour. The creation of mindful spaces which allow for meaningful engagement is particularly relevant for Gen Z consumers. As a generation with the greatest exposure to media and constant free-flowing information, opportunities to be content in the present are a particular challenge for this group (Bhalla et al., 2021).

An additional element of personalisation is interaction with staff. However, with the growth in digital resources, these interactions within the servicescape are changing considerably. A key area for consideration is the growth in Artificial Intelligence (AI) and robots within service spaces. AI chatbots for example are a firm feature in tourism and hospitality communications. As discussed by Whang et al. (2022), as AI technology and the availability of big data have grown, the level of personalisation in chatbot communications has increased. As such, customers can interact more naturally with chatbots and receive tailored information, guidance, and support in the service encounter. Equally, the growth in anthropomorphic (human-like) robotics which simulates human characteristics provides opportunities for customers to have personalised contact with a non-human entity. This, however, requires thought and management as customer responses to robotic interactions can be met with uncertainty and anxiety. This also has the potential to create a barrier to meaningful interactions between customers and staff (Xie & Lei, 2022). As such, the hybrid servicescape needs to capitalise on the benefits of digital tools to support personalisation but equally consider how these will complement (or conflict) with the physical and ambient conditions.

Customer Control

In line with underpinning principles of co-creation, that customers are actively involved in creating their own experience and therefore the subsequent value derived from it requires a degree of customer control (Hamidi et al., 2020). As discussed by Shulga and Busser (2021), customer self-determination and control over personal actions and interactions have strong effects on customer well-being, satisfaction, and experience. As such, customer autonomy and control have increasingly become part of the service encounter and a feature of the hybrid servicescape. This can particularly be seen in the self-service technology (SST) within tourism and hospitality settings. Safaeimanesh et al. (2021) highlight the power that SST has for various customer needs (such as enjoyment, convenience, and customisation) but also note that there are varying preferences for SST as a leisure vs. business customer. From a servicescape perspective, designing a space that provides features where customers can exercise control and autonomy requires both physical and digital tools. Self-service applications within the space, such as fixed autonomous points in airports, visitor attractions, and hospitality venues, afford customers with additional discretion and privacy.

Aside from SST, the hybrid servicescape can provide additional opportunities for customer control. At a physical level, there are various ways that customers could self-determine the experience they engage with. Choosing preferred visitor routes in natural attractions through a variety of trails at a small level exerts control. Equally, in museums, galleries, and visitor centres, providing a range of tangible interpretation, such as printed material, storyboards, and replica, allows customers to select how they wish to engage with the underpinning narrative of the site (Urquhart & Leask, 2019). Equally, incorporating digital resources which encourage customers to take ownership of the experience they engage with can act as a driver for co-creation (Kumar & Vinitha, 2020). As discussed within the Internet of Things (IoT), contemporary customers in tourism and hospitality can now use interconnected systems to have greater autonomy in travel booking and purchasing. Within the hybrid servicescape, a combination of physical and ambient tools in addition to interconnected digital resources provide the scope for seamless experience design that champions customer control and autonomy. Examples of this were observed in the theme park sector post-Covid-19 with the use of virtual queuing applications to replace long waiting times. Such an approach not only empowers visitors to take control over the planned visit and schedule but also acts to enhance the experience for younger visitors by removing unnecessary waiting for rides and attractions (Bonfanti et al., 2023).

Conclusions

For tourism and hospitality businesses, the hybrid servicescape is already present in various environments. However, with enhanced digital resources and the constant growth of digital capabilities within the physical space, there is an increasing need to reframe our view of the experience environment. As such, there are opportunities to explore customer responses and behaviours not only in relation to physical and ambient evidence but also in relation to digital touchpoints as part of the experience. From a management perspective, this requires strategic thought as to the role and value posed by digital tools that complement the physical and ambient space. As explored throughout the chapter, there are numerous opportunities to integrate digital touchpoints throughout the customer journey that can (in combination with physical resources) create a co-creative environment. Future service designers will continue to have a vast toolkit of physical, ambient, and digital resources to choose from, however, a key challenge will be to design these into a cohesive and seamless space that fosters active dialogue, engagement, personalisation, and customer control.

Recommended Further Reading

Allmér, H. R. (2020). Servicescape is. In M. Helmefalk & L. Marcusson (Eds.), *Utilizing gamification in servicescapes for improved consumer engagement* (pp. 1–23). IGI Global. https://doi.org/10.4018/978-1-7998-1970-7.ch001

Bitner, M. J. (1992). Servicescapes – The impact of physical surroundings on customers and employees. *Journal of Marketing, 56*(2), 57–71. https://doi.org/10.2307/1252042

Heo, J., Lee, W., & Moon, J. (2023). Structural relationship between theme park servicescape, instagramability, brand attitude and intention to revisit. *Sustainability, 15*(13), 9935. https://doi.org/10.3390/su15139935

Line, N., & Hanks, L. (2019). The social servicescape: A multidimensional operationalization. *Journal of Hospitality and Tourism Research, 43*(2), 167–187. https://doi.org/10.1177/1096348018767948

Pizam, A., & Tasci, A. D. (2019). Experiencescape: Expanding the concept of servicescape with a multi-stakeholder and multi-disciplinary approach. *International Journal of Hospitality Management, 76*, 25–37. https://doi.org/10.1016/j.ijhm.2018.06.010

Rosenbaum, M., & Massiah, C. (2011). An expanded servicescape perspective. *Journal of Service Management, 22*(4), 471–490. https://doi.org/10.1108/09564231111155088

Roy, S., Singh, G., Hope, M., Nguyen, B., & Harrigan, P. (2019). The rise of smart consumers: Role of smart servicescape and smart consumer experience co-creation. *Journal of Marketing Management, 35*(15–16), 1480–1513. https://doi.org/10.1080/0267257X.2019.1680569

References

Allmér, H. R. (2020). Servicescape is. In M. Helmefalk & L. Marcusson (Eds.), *Utilizing gamification in servicescapes for improved consumer engagement* (pp. 1–23). IGI Global. https://doi.org/10.4018/978-1-7998-1970-7.ch001

Alyahya, M., & McLean, G. (2022). Examining tourism consumers' attitudes and the role of sensory information in virtual reality experiences of a tourist destination. *Journal of Travel Research, 61*(7), 1666–1681. https://doi.org/10.1177/00472875211037745

Amitrano, C. C., Russo Spena, T., & Bifulco, F. (2021). Augmented servicescape: Integrating physical and digital reality. In T. Russo Spena & F. Bifulco (Eds.), *Digital transformation in the cultural heritage sector. Contributions to management science*. Springer. https://doi.org/10.1007/978-3-030-63376-9_9

Antwi, C. O., Ren, J., Zhang, W., Owusu-Ansah, W., Aboagye, M. O., Affum-Osei, E., & Agyapong, R. A. (2022). "I Am Here to Fly, but Better Get the Environment Right!" Passenger response to airport servicescape. *Sustainability, 14*(16), 10114. https://doi.org/10.3390/su141610114

Apaolaza, V., Hartmann, P., Fernández-Robin, C., & Yáñez, D. (2020). Natural plants in hospitality servicescapes: The role of perceived aesthetic value. *International Journal of Contemporary Hospitality Management, 32*(2), 665–682. https://doi.org/10.1108/IJCHM-03-2019-0240

Bagchi, S., He, J., Chen, C. Y. F., Zhang, M., & Bhati, A. (2021). Creating excellent guest experiences: Servicescape and processes. In K. Thirumaran, D. Klimkeit, & C. M. Tang (Eds.), *Service excellence in tourism and hospitality: Insights from Asia. Springer International Publishing.* https://doi.org/10.1007/978-3-030-57694-3

Bhalla, R., Tiwari, P., & Chowdhary, N. (2021). Digital natives leading the world: Paragons and values of generation Z. In N. Stylos, R. Rahimi, B. Okumus, & S. Williams (Eds.), *Generation Z marketing and management in tourism and hospitality* (pp. 3–23). Palgrave Macmillan. https://doi.org/10.1007/978-3-030-70695-1_1

Bilgili, B., Ozkul, E., & Koc, E. (2020). The influence of colour of lighting on customers' waiting time perceptions. *Total Quality Management & Business Excellence, 31*(9–10), 1098–1111. https://doi.org/10.1080/14783363.2018.1465814

Bitner, M. J. (1992). Servicescapes – The impact of physical surroundings on customers and employees. *Journal of Marketing, 56*(2), 57–71. https://doi.org/10.2307/1252042

Bonfanti, A., Vigolo, V., Gonzo, E., & Genuardi, I. (2023). Customer experience management in themed amusement parks: The impact of safety measures on the physical, social and digital servicescape of Gardaland Park. *Consumer Behavior in Tourism and Hospitality*. https://doi.org/10.1108/CBTH-04-2022-0099

Buhalis, D., & Sinarta, Y. (2019). Real-time co-creation and nowness service: Lessons from tourism and hospitality. *Journal of Travel & Tourism Marketing, 36*(5), 563–582. https://doi.org/10.1080/10548408.2019.1592059

Bush, E. (2022). New luxury travel agency features virtual reality cinema room. *Travel Weekly*. https://travelweekly.co.uk/news/travel-agents/new-luxury-travel-agency-features-virtual-reality-cinema-room

Chapman, D. (2017). How to use the Internet of Things to elevate the customer experience. *Hotel Management*. https://www.hotelmanagement.net/tech/how-to-use-internet-things-to-elevate-customer-experience

Chathoth, P. K., Ungson, G. R., Harrington, R. J., & Chan, E. S. W. (2016). Co-creation and higher order customer engagement in hospitality and tourism services: A critical review. *International Journal of Contemporary Hospitality Management, 28*(2), 222–245. https://doi.org/10.1108/IJCHM-10-2014-0526

Choy, M. W. C., & Shih, C. C. C. (2022). The effect of employee uniform on job satisfaction: A case of the housekeeping department in a luxury five-star hotel in Hong Kong, China. *Tourism and Hospitality Management, 28*(3), 559–574. https://doi.org/10.20867/thm.28.3.5

Errajaa, K., Legohérel, P., Daucé, B., & Bilgihan, A. (2021). Scent marketing: Linking the scent congruence with brand image. *International Journal of Contemporary Hospitality Management, 33*(2), 402–427. https://doi.org/10.1108/IJCHM-06-2020-0637

Evans, G. (2012). Hold back the night: Nuit Blanche and all-night events in capital cities. *Current Issues in Tourism, 15*(1–2), 35–49. https://doi.org/10.1080/13683500.2011.634893

Fox, J. (2018). Smart mirrors are helping redefine the hotel guest experience. *Hotel Management*. https://www.hotelmanagement.net/tech/how-smart-mirrors-are-changing-way-guests-interact-hotel-rooms

Goh, E., & Lee, C. (2018). A workforce to be reckoned with: The emerging pivotal Generation Z hospitality workforce. *International Journal of Hospitality Management, 73*, 20–28. https://doi.org/10.1016/j.ijhm.2018.01.016

Gremyr, I., Birch-Jensen, A., Kumar, M., & Löfberg, N. (2022). Quality functions' use of customer feedback as activation triggers for absorptive capacity

and value co-creation. *International Journal of Operations & Production Management, 42*(13), 218–242. https://doi.org/10.1108/IJOPM-11-2021-0692
Grönroos, C., & Voima, P. (2013). Critical service logic: Making sense of value creation and co-creation. *Journal of the Academy of Marketing Science, 41*(2), 133–150. https://doi.org/10.1007/s11747-012-0308-3
Gursoy, D., Malodia, S., & Dhir, A. (2022). The metaverse in the hospitality and tourism industry: An overview of current trends and future research directions. *Journal of Hospitality Marketing & Management, 31*(5), 527–534. https://doi.org/10.1080/19368623.2022.2072504
Hamidi, F., Gharneh, N. S., & Khajeheian, D. (2020). A conceptual framework for value co-creation in service enterprises (case of tourism agencies). *Sustainability, 12*(1), 213. https://doi.org/10.3390/su12010213
Hammady, R., Ma, M., AL-Kalha, Z., & Strathearn, C. (2021). A framework for constructing and evaluating the role of MR as a holographic virtual guide in museums. *Virtual Reality: The Journal of the Virtual Reality Society, 25*(4), 895–918. https://doi.org/10.1007/s10055-020-00497-9
Han, H. (2013). Effects of in-flight ambience and space/function on air travelers' decision to select a low-cost airline. *Tourism Management, 37*, 125–135. https://doi.org/10.1016/j.tourman.2013.01.008
Han, H., & Ryu, K. (2009). The roles of the physical environment, price perception, and customer satisfaction in determining customer loyalty in the restaurant industry. *Journal of Hospitality and Tourism Research, 33*(4), 487–510. https://doi.org/10.1177/1096348009344212
Haobin, Y. B., Huiyue, Y., Peng, L., & Fong, L. H. N. (2021). The impact of hotel servicescape on customer mindfulness and brand experience: The moderating role of length of stay. *Journal of Hospitality Marketing & Management, 30*(5), 592–610. https://doi.org/10.1080/19368623.2021.1870186
Harris, L., & Goode, M. (2010). Online servicescapes, trust, and purchase intentions. *The Journal of Services Marketing, 24*(3), 230–243. https://doi.org/10.1108/08876041011040631
Heo, J., Lee, W., & Moon, J. (2023). Structural relationship between theme park servicescape, instagramability, brand attitude and intention to revisit. *Sustainability, 15*(13), 9935. https://doi.org/10.3390/su15139935
Hussain, Z. (2021). Paradigm of technological convergence and digital transformation: The challenges of CH sectors in the global COVID-19 pandemic and commencing resilience- based structure for the post-COVID-19 era. *Digital Applications in Archaeology and Cultural Heritage, 21*, e00182–e00182. https://doi.org/10.1016/j.daach.2021.e00182

Iskender, A., Sirakaya-Turk, E., Cardenas, D., & Harrill, R. (2022). COVID or VOID: A systematic literature review of technology adoption and acceptance in hospitality and tourism since the breakout of COVID-19. *Tourism and Hospitality Research*. https://doi.org/10.1177/14673584221133667

Ivanov, S., Gretzel, U., Berezina, K., Sigala, M., & Webster, C. (2019). Progress on robotics in hospitality and tourism: A review of the literature. *Journal of Hospitality and Tourism Technology, 10*(4), 489–521. https://doi.org/10.1108/JHTT-08-2018-0087

Jain, G., Paul, J., & Shrivastava, A. (2021). Hyper-personalization, co-creation, digital clienteling and transformation. *Journal of Business Research, 124*, 12–23. https://doi.org/10.1016/j.jbusres.2020.11.034

Jang, Y. J. (2021). The role of customer familiarity in evaluating green servicescape: An investigation in the coffee shop context. *International Journal of Contemporary Hospitality Management, 33*(2), 693–716. https://doi.org/10.1108/IJCHM-04-2020-0356

Juma, L., Bakos, I. M., & Khademi-Vidra, A. (2020). Nature interpretation and visitor management objectives: A survey of tourist attitudes at Maasai Mara National Reserve, Kenya. *Sustainability, 12*(18), 7246. https://doi.org/10.3390/su12187246

Kazlauskaite, R., Martinaityte, I., Lyubovnikova, J., & Augutyte-Kvedaraviciene, I. (2022). The physical office work environment and employee wellbeing: Current state of research and future research agenda. *International Journal of Management Reviews: IJMR*. https://doi.org/10.1111/ijmr.12315

Kelly, P., & Lawlor, J. (2021). Adding or destroying value? User experiences of tourism self- service technologies. *Journal of Hospitality and Tourism Insights, 4*(3), 300–317. https://doi.org/10.1108/JHTI-08-2018-0051

Kim, M., & Baker, M. A. (2022). From surviving to co-creating: the effects of the reshaped physical and social servicescape on customer citizenship behavior. *The Service Industries Journal*, 1–26. https://doi.org/10.1080/02642069.2022.2147513

Kim, S., Park, E., & Xu, M. (2020). Beyond the authentic taste: The tourist experience at a food museum restaurant. *Tourism Management Perspectives, 36*, 100749. https://doi.org/10.1016/j.tmp.2020.100749

Kirillova, K., Fu, X., & Kucukusta, D. (2020). Workplace design and well-being: Aesthetic perceptions of hotel employees. *The Service Industries Journal: Well-Being Research in the Service Industries, 40*(1–2), 27–49. https://doi.org/10.1080/02642069.2018.1543411

Kumar, D. S., & Vinitha, V. U. (2020). Servicescapes and E-servicescapes design issues in emerging markets: Imperatives, challenges and agenda setting. In A. Adhikari (Ed.), *Services marketing issues in emerging economies*. Springer. https://doi.org/10.1007/978-981-15-8787-0_1

Lai, K. P., & Chong, S. C. (2020). The influence of servicescape and service credibility on older adults' intention to recover – A study of rehabilitation services in Malaysia. *Journal of Health Organization and Management, 34*(2), 101–122. https://doi.org/10.1108/JHOM-07-2019-0215

Lee, S. (2018). m-servicescape: Effects of the hotel mobile app servicescape preferences on customer response. *Journal of Hospitality and Tourism Technology, 9*(2), 172–187. https://doi.org/10.1108/JHTT-08-2017-0066

Lee, S., & Chuang, N.-K. (2022). Applying expanded servicescape to the hotel industry. *Journal of Hospitality and Tourism Research, 46*(4), 771–796. https://doi.org/10.1177/10963480211011535

Lee, W., & Kim, Y. (2021). Does VR Tourism Enhance Users' Experience? *Sustainability, 13*(2), 806. https://doi.org/10.3390/su13020806

Lei, S. I., Wang, D., & Law, R. (2019). Hoteliers' service design for mobile-based value co- creation. *International Journal of Contemporary Hospitality Management, 31*(11), 4338–4356. https://doi.org/10.1108/IJCHM-03-2018-0249

Lestari, S. M., Azizah, G. N., Sari, D. P., & Maulana, H. (2021). The impact of COVID-19 pandemic on café design concept. *International Journal of Research and Applied Technology, 1*(2), 431–438. https://doi.org/10.34010/injuratech.v1i2.6778

Lin, Y.-C., Lai, H.-J., & Morrison, A. M. (2019). Social servicescape and Asian students: An analysis of spring break island bed and breakfast experiences in Taiwan. *Tourism Management Perspectives, 31*, 165–173. https://doi.org/10.1016/j.tmp.2019.04.005

Lin, I. Y., & Mattila, A. S. (2010). Restaurant servicescape, service encounter, and perceived congruency on customers' emotions and satisfaction. *Journal of Hospitality Marketing & Management, 19*(8), 819–841. https://doi.org/10.1080/19368623.2010.514547

Lin, I. Y., & Worthley, R. (2012). Servicescape moderation on personality traits, emotions, satisfaction, and behaviors. *International Journal of Hospitality Management, 31*(1), 31–42. https://doi.org/10.1016/j.ijhm.2011.05.009

Line, N., & Hanks, L. (2019). The social servicescape: A multidimensional operationalization. *Journal of Hospitality and Tourism Research, 43*(2), 167–187. https://doi.org/10.1177/1096348018767948

Liu, S. Q., Bogicevic, V., & Mattila, A. S. (2018). Circular vs. angular servicescape: "Shaping" customer response to a fast service encounter pace. *Journal of Business Research, 89*, 47–56. https://doi.org/10.1016/j.jbusres.2018.04.007

Lu, V. N., Wirtz, J., Kunz, W. H., Paluch, S., Gruber, T., Martins, A., & Patterson, P. G. (2020). Service robots, customers and service employees: What can we learn from the academic literature and where are the gaps? *Journal of Service Theory and Practice, 30*(3), 361–391. https://doi.org/10.1108/JSTP-04-2019-0088

Lu, Y.-H., Zhang, J., Zhang, H., Xiao, X., Liu, P., Zhuang, M., & Hu, M. (2022). Flow in soundscape: The conceptualization of soundscape flow experience and its relationship with soundscape perception and behavior intention in tourism destinations. *Current Issues in Tourism, 25*(13), 2090–2108. https://doi.org/10.1080/13683500.2021.1922363

MacLeod, N. (2017). The role of trails in the creation of tourist space. *Journal of Heritage Tourism, 12*(5), 423–430. https://doi.org/10.1080/1743873X.2016.1242590

Madzharov, A. (2022). Scents research and its applications in tourism. *Annals of Tourism Research, 93*, 103309. https://doi.org/10.1016/j.annals.2021.103309

Manthiou, A., Klaus, P., Kuppelwieser, V. G., & Reeves, W. (2021). Man vs machine: Examining the three themes of service robotics in tourism and hospitality. *Electronic Markets, 31*(3), 511–527. https://doi.org/10.1007/s12525-020-00434-3

Massari, F. S., Del Vecchio, P., & Degl'innocenti, E. (2022). Past for future – Museums as a digitalized "interaction platform" for value co-creation in tourism destinations. *European Journal of Innovation Management*. https://doi.org/10.1108/EJIM-09-2022-0521

Mehrabian, A., & Russell, J. A. (1976). *An approach to environmental psychology*. MIT Press.

Merdin-Uygur, E., & Ozturkcan, S. (2023). Consumers and service robots: Power relationships amid COVID-19 pandemic. *Journal of Retailing and Consumer Services, 70*, 103174. https://doi.org/10.1016/j.jretconser.2022.103174

Novak, C. C., La Lopa, J., & Novak, R. E. (2010). Effects of sound pressure levels and sensitivity to noise on mood and behavioral intent in a controlled fine dining restaurant environment. *Journal of Culinary Science & Technology, 8*(4), 191–218. https://doi.org/10.1080/15428052.2010.535756

Nyrhinen, J., Uusitalo, O., Frank, L., & Wilska, T.-A. (2022). How is social capital formed across the digital-physical servicescape? *Digital Business, 2*(2), 100047. https://doi.org/10.1016/j.digbus.2022.100047

Oyner, O., & Korelina, A. (2016). The influence of customer engagement in value co- creation on customer satisfaction. *Worldwide Hospitality and Tourism Themes, 8*(3), 327–345.

Peng, Y., Wang, E. Y., & Lam, D. (2022). Build it and they will come? The impact of servicescape on Chinese millennials' satisfaction and behavioral intentions toward integrated resorts. *International Journal of Hospitality & Tourism Administration, 23*(3), 576–598. https://doi.org/10.1080/15256480.2020.1805090

Pizam, A., & Tasci, A. D. (2019). Experiencescape: Expanding the concept of servicescape with a multi-stakeholder and multi-disciplinary approach. *International Journal of Hospitality Management, 76*, 25–37. https://doi.org/10.1016/j.ijhm.2018.06.010

Prayag, G., Gannon, M. J., Muskat, B., & Taheri, B. (2020). A serious leisure perspective of culinary tourism co-creation: The influence of prior knowledge, physical environment and service quality. *International Journal of Contemporary Hospitality Management, 32*(7), 2453–2472. https://doi.org/10.1108/IJCHM-10-2019-0897

Ramaswamy, V., & Ozcan, K. (2018). What is co-creation? An interactional creation framework and its implications for value creation. *Journal of Business Research, 84*, 196–205. https://doi.org/10.1016/j.jbusres.2017.11.027

Ramaswamy, V., & Ozcan, K. (2020). The "Interacted" actor in platformed networks: Theorizing practices of managerial experience value co-creation. *The Journal of Business & Industrial Marketing, 35*(7), 1165–1178. https://doi.org/10.1108/JBIM-10-2018-0318

Rihova, I., Buhalis, D., Gouthro, M. B., & Moital, M. (2018). Customer-to-customer co- creation practices in tourism: Lessons from customer-dominant logic. *Tourism Management, 67*, 362–375. https://doi.org/10.1016/j.tourman.2018.02.010

Ristova, C., & Dimitrov, N. (2019). Digitalization in the hospitality industry: Trends that might shape the next stay of guests. *International Journal of Information, Business and Management, 11*(3), 144–154. https://www.proquest.com/docview/2216865195/fulltext/48D4996178B0411DPQ/1?accountid=16607

Rosenbaum, M., & Massiah, C. (2011). An expanded servicescape perspective. *Journal of Service Management, 22*(4), 471–490. https://doi.org/10.1108/09564231111155088

Roy, S., Singh, G., Hope, M., Nguyen, B., & Harrigan, P. (2019). The rise of smart consumers: Role of smart servicescape and smart consumer experience co-creation. *Journal of Marketing Management, 35*(15–16), 1480–1513.

Ryu, Y. K., & Park, J.-W. (2019). Investigating the Effect of Experience in an Airport on Pleasure Satisfaction and Airport Image: A Case Study on Incheon International Airport. *Sustainability, 11*(17). https://doi.org/4616-10.3390/su11174616
Safaeimanesh, F., Kilic, H., Alipour, H., & Safaeimanesh, S. (2021). Self-service technologies (SSTs) – The next frontier in service excellence: Implications for tourism industry. *Sustainability, 13*(5), 2604. https://doi.org/10.3390/su13052604
Serrano-Baena, M. M. M., Fernandez, R. E. H., Ruiz-Diaz, C., & Trivino-Tarradas, P. (2023). Promoting the sustainable recovery of hospitality in the post-pandemic era: A comparative study to optimize the servicescapes. *International Journal of Environmental Research and Public Health, 20*(2), 1100. https://doi.org/10.3390/ijerph20021100
Sfandla, C., & Björk, P. (2013). Tourism experience network: Co-creation of experiences in interactive processes. *The International Journal of Tourism Research, 15*(5), 495–506. https://doi.org/10.1002/jtr.1892
Shulga, L. V., & Busser, J. A. (2021). Customer self-determination in value co-creation. *Journal of Service Theory and Practice, 31*(1), 83–111.
Tankovic, A., & Benazic, D. (2018). The perception of e-servicescape and its influence on perceived e-shopping value and customer loyalty. *Online Information Review, 42*(7), 1124–1145. https://doi.org/10.1108/OIR-12-2016-0354
Tarlao, C., Fernandez, P., Frissen, I., & Guastavino, C. (2021). Influence of sound level on diners' perceptions and behavior in a Montreal restaurant. *Applied Acoustics, 174*, 107772. https://doi.org/10.1016/j.apacoust.2020.107772
Tasci, A. D., & Pizam, A. (2020). An expanded nomological network of experiencescape. *International Journal of Contemporary Hospitality Management, 32*(3), 999–1040. https://doi.org/10.1108/IJCHM-12-2018-0988
Tran Xuan, Q., Dang Van, M., & Tournois, N. (2022). The effects of café servicescape on employee job satisfaction-centered on social interaction, job experience and emotional experience: A study in Vietnam. *Journal of Human Resources in Hospitality & Tourism, 21*(2), 197–225. https://doi.org/10.1080/15332845.2022.2031607
Urquhart, E. (2019). Technological mediation in the future of experiential tourism. *Journal of Tourism Futures, 5*(2), 120–126. https://doi.org/10.1108/JTF-04-2019-0033

Urquhart, E. (2022). Interpretative tools. In D. Buhalis (Ed.), *Encyclopaedia of tourism management and marketing* (pp. 779–782). Edward Elgar Publishing.

Urquhart, E., & Leask, A. (2019). Managing the co-creation of narratives in the heritage sector: The case of the Surgeons' Hall Museum, Edinburgh. In A. Correia, A. Fyall, & M. Kozak (Eds.), *Experiential consumption and marketing in tourism with a cross- cultural context* (pp. 157–173). Goodfellow Publishers.

Utkarsh, & Sigala, M. (2021). A bibliometric review of research on COVID-19 and tourism: Reflections for moving forward. *Tourism Management Perspectives, 40*, 100912–100912. https://doi.org/10.1016/j.tmp.2021.100912

Vargo, S., & Lusch, R. (2004). Evolving to a new dominant logic for marketing. *Journal of Marketing, 68*(1), 1–17.

Verhoeven, J. W. M., van Rompay, T. J. L., & Pruyn, A. T. H. (2009). The price facade: Symbolic and behavioral price cues in service environments. *International Journal of Hospitality Management, 28*(4), 604–611. https://doi.org/10.1016/j.ijhm.2009.03.013

Vuksanovic, N., Bajrami, D. D., Petrovic, M. D., & Grigorieva, E. M. (2021). QR codes as a tool for receiving feedback about guests' satisfaction at destinations. *Journal of Place Management and Development, 14*(1), 19–31. https://doi.org/10.1108/JPMD-03-2020-0025

Wan, Y. K. P. (2022). Accessibility of tourist signage at heritage sites: An application of the universal design principles. *Tourism Recreation Research*, 1–15. https://doi.org/10.1080/02508281.2022.2106099

Wei, W. (2019). Research progress on virtual reality (VR) and augmented reality (AR) in tourism and hospitality: A critical review of publications from 2000 to 2018. *Journal of Hospitality and Tourism Technology, 10*(4), 539–570. https://doi.org/10.1108/JHTT-04-2018-0030

Whang, J.-B., Song, J. H., Lee, J.-H., & Choi, B. (2022). Interacting with Chatbots: Message type and consumers' control. *Journal of Business Research, 153*, 309–318. https://doi.org/10.1016/j.jbusres.2022.08.012

Xie, L., & Lei, S. (2022). The nonlinear effect of service robot anthropomorphism on customers' usage intention: A privacy calculus perspective. *International Journal of Hospitality Management, 107*, 103312. https://doi.org/10.1016/j.ijhm.2022.103312

Yen, C.-H., Teng, H.-Y., & Tzeng, J.-C. (2020). Innovativeness and customer value co-creation behaviors: Mediating role of customer engagement. *International Journal of Hospitality Management, 88*, 102514. https://doi.org/10.1016/j.ijhm.2020.102514

Yin, C. Z. Y., Jung, T., Tom Dieck, M. C., & Lee, M. Y. (2021). Mobile augmented reality heritage applications: Meeting the needs of heritage tourists. *Sustainability, 13*(5), 2523. https://doi.org/10.3390/su13052523

Yin, D., Li, M., Qiu, H., Bai, B., & Zhou, L. (2023). When the servicescape becomes intelligent: Conceptualization, assessment, and implications for hospitableness. *Journal of Hospitality and Tourism Management, 54*, 290–299. https://doi.org/10.1016/j.jhtm.2023.01.001

Yung, R., & Khoo-Lattimore, C. (2019). New realities: A systematic literature review on virtual reality and augmented reality in tourism research. *Current Issues in Tourism, 22*(17), 2056–2081. https://doi.org/10.1080/13683500.2017.1417359

Zhang, T., Li, B., Huang, A., & Hua, N. (2023). Examining a perceived value model of servicescape for bed-and-breakfasts. *Journal of Quality Assurance in Hospitality & Tourism, 24*(4), 359–379. https://doi.org/10.1080/1528008X.2022.2051219

Zheng, D., Liu, S., & Lu, W. (2022). Do you trust digital health pass? Understanding tourists' responses toward using health QR codes in pandemic travel. *Journal of China Tourism Research*, 1–19. https://doi.org/10.1080/19388160.2022.2087817

6

Creativity and Marketing with New Technologies in Tourism and Hospitality

Orhan Can Yılmazdoğan, Cemile Ece, and Efnan Ezenel

Introduction

Tourism is one of the many fields where creativity has been used, and it has been used from the past to the present in a variety of fields in an attempt to create innovative ways. Creativity and creative thinking have been incorporated into the marketing of the tourism business, especially for increasing market competitiveness and reaching customers (Sofronov, 2019). Technology development has increased the prominence of creative marketing techniques, and the emergence of the digital era has also resulted in adjustments to these strategies. Companies are employing novel digital marketing strategies and tools, particularly in the connection between social media and Internet-based applications, to reach new target markets and improve their market share (Mkwizu, 2019).

Tourism marketing strategies have evolved significantly over time with the integration of digital marketing and creative applications. Innovative

O. C. Yılmazdoğan (✉) • C. Ece • E. Ezenel
Eskişehir Osmangazi University, Eskişehir, Turkey
e-mail: cyilmazdogan@ogu.edu.tr

N. Stylos et al. (eds.), *Contemporary Marketing Management for Tourism and Hospitality*, https://doi.org/10.1007/978-3-031-65049-9_6

digital marketing techniques are constantly being added to the equation, along with evolving technologies that favor creativity. Recent technological advances, particularly the development of artificial intelligence, have led to the emergence of new tools and applications that can combine marketing techniques by developing various innovative concepts. Creative techniques have emerged from developments in digital marketing. This has led to the adoption of these principles as standard practices in other industries, including tourism (Yoga et al., 2019).

In this chapter, in-depth knowledge about creativity, digital marketing, the development of new technologies, and their implications for tourism has been gathered in order to study their contribution to creativity and marketing within the process of digitization in the tourism industry.

Creativity

All jobs and professions, both intellectual and emotional, have been influenced by the concept of creativity from ancient times. As stated by Young (1985), creative processes include finding new topics, producing unique products and cognitive processes, resolving problems, and creating value. The structure of the creativity process varies depending on the discipline because it comprises novel, innovative, and distinctive processes and products that could emerge in a certain industry as a result of learning, reasoning, and communication. There are, however, variations and new productions based on it (Rhodes, 1961; Henriksen et al., 2014).

Regardless of the different studies and definitions of creativity, it is obvious that the desire to be creative in all its forms and bases—which is fundamentally the desire to generate something of value—is what creativity is all about. The desire to communicate a message and make an influence on the target market or community that the product or process reaches can be a value-creation process in creativity. Value creation can be assessed in terms of fostering a specific culture's consciousness or supporting a specific ideology (Dollinger et al., 2007). Therefore, invention and creativity are sources of these techniques.

According to Ludwig (1989), the preparation stage is the first step in the creative process. This stage defines the problem, highlights the available options and what can be done to address the problem, and formulates the problem. Following that, the problem is placed on wait, and the incubation phase begins. At this point, when people are preoccupied with something else, ideas usually begin to form. Following it, enlightenment and discovery occur. In line with social, artistic, and/or scientific norms, the knowledge that is currently available is evaluated, and conceptions are then tested and combined. At this point, an attempt is made to identify striking parallels between the ideas, to establish previously unrecognized relationships between them, and then to integrate them into a new whole. In other words, the process of creating a new idea or concept involves discovering new relationships and insights that weren't apparent before.

The relationship between the creative person, process, and outcome feeds the phenomena of creativity. Each of these three defining categories highlights different aspects and dynamics of creativity. It is noted that creativity is an inborn phenomenon that can be developed when definitions of creativity related to creative personality attributes are assessed (Kienitz et al., 2014). By forming many connections to yield different solutions, it is the goal of the creative process to reveal something novel. Creativity produces both the resurrection of long-established realities and the emergence of new realities. In definitions of creativity that are related to the creative product, the emphasis is on the innovative and unique idea (Onur & Zorlu, 2017).

The fact that creativity occurs in every sector and discipline increases competitiveness. Particularly in organizations, many strategies and approaches are developed along with creativity to increase market share. It is extremely important to create the necessary business environment for the development of an understanding of innovation in organizations and the use of creative talents working with managerial mechanisms to support this. In this way, drawing can enhance creativity and develop innovative ideas (Yahyagil, 2001).

The Importance of Creativity for Digital Marketing

Through innovation, whether by changing the demand or cost curves for their products, or by gaining a competitive advantage by enhancing their capacity to innovate, organizations seek to improve performance. To maintain and strengthen their position in the market, to take advantage of the opportunities presented by change, to overcome threats, to increase the demand for products/brands, to have a say in their sectors, to grow, and to find solutions to marketing problems, innovative organizations that are open to innovations and support new ideas implement marketing innovation strategies (Matejun, 2018). Employees who tend to solve the marketing issues brought on by change and uncertainty and who present helpful, novel, and innovative solutions can help their firms create value. To remain in business, to safeguard or solidify their market positions, to avoid falling behind innovations provided by rivals, to capitalize on opportunities presented by change, to avoid threats, to reach consumers with shifting expectations and demands faster and more effectively than rivals, and to boost demand for new and/or existing products. Marketing innovation strategies are becoming more and more significant (McIntyre, 1993).

For businesses to be able to handle the escalating competition and avoid uncertainty, creativity is crucial. Because highly effective marketing strategies and tactics that foster a sense of value between businesses and their customers, as well as the development of solutions and recommendations for their requirements and wants, are all part of creative practices. It has been noted that marketing that emphasizes creativity yields better and more effective results (Kumar, 2018). Different applications have emerged in this day, where marketing and creativity cannot be seen as separate concepts, particularly with the advancement of technology. It has been noted that my organizations have started to develop digitally and incorporate digital applications and marketing strategies in order to take advantage of the market and boost their competitive power given the role that technology plays in human life (Das et al., 2023).

When the relationship between digital marketing and creativity is examined, it is necessary to have a technological infrastructure and a social network in order for companies to make their lives sustainable and to gain a solid position in the market or for a newly opened company to create a market of its own, especially with technology being in such a large human life. Thanks to social networks and digital marketing strategy, businesses can reach customers more easily and work with less cost. Therefore, the use of developing technology plays an effective role, especially in content production and value-creation processes between customers and companies. The use of creative applications in digital marketing contributes both to the production of more comfortable content using technology and to the creation of a sustainable marketing strategy for companies (Dora & Saudi, 2020). Digitalization, especially in marketing, serves as an automation in collecting information about customers, analyzing them easily, performing the company's Life Cycle marketing, and creating promotional and marketing content with very little effort. Especially thanks to Artificial Intelligence, companies can analyze big data and take necessary steps by developing creative marketing strategies for them (Stylos et al., 2021; Stylos & Zwiegelaar, 2019).

In this period when digitalization is extremely important and is at the forefront in every field, digital marketing provides great advantages to businesses when combined with creative methods and can bring businesses to the desired level of recognition. Since the coexistence of digital applications and creativity in marketing will provide companies with a great advantage, companies need to transform their traditional marketing approach into a digital marketing approach by combining them with modern technology.

Digital Creativity

Today's economy is changing the way we create, communicate, work, and interact. Creativity is now one of the main drivers of today's digital business environment. In the new digital era, creativity and innovation are key to creating value for businesses (Sousa & Rocha, 2019). Businesses go through the process of digital transformation by developing new digital business models that help to create and capture value, as well as gradually changing the way they use digital technologies (Smailhodžić & Berberović, 2021). Consequently, businesses are increasingly looking for innovative individuals who can provide fresh perspectives on old challenges. Digital creativity—the use of digital tools and technologies to explore creative ideas and innovative methods to exhibit your ideas, research, or work—is becoming increasingly recognized as a necessary ability for graduates from a wide range of fields (Janse van Rensburg et al., 2022). Accordingly, the concept of digital creativity has emerged in businesses and all forms of creativity that are influenced by digital technologies are referred to as "digital creativity" (Lee & Chen, 2015). Achievement in digital creativity requires the highest level of technical competence and understanding and implies a specific level of acknowledged achievement (e.g., valued-based certification, winning an award, confirmed diplomas) (Hoffmann et al., 2016).

This clearly shows that digital creativity and digital intelligence are linked. It is therefore necessary to have a description of the concept of digital intelligence at the same time as a description of digital creativity. Broadly speaking, digital intelligence is the ability to learn and use new information and skills related to digital technologies such as social, mobile, analytics, cloud, and, more recently, Internet security (Battro, 2009). In business, however, digital intelligence is more than just the ability to use digital technologies. It addresses what, why, where, when, who, how, and how much of digital technology to improve operational efficiency and outcomes (Boughzala et al., 2020; Waller, 2020). Digital intelligence encompasses the knowledge of how to manage IT, coordinate business and IT strategies, and manage IT projects and enterprise systems (Mithas & McFarlan, 2017). It goes beyond the simple knowledge of how to use computers or other forms of IT. Consequently, our interaction with technology and the impact of digital creativity are fundamentally related to digital intelligence. Therefore, digital creativity is the result of a creative process that has the use of computational technology.

For tourism destinations, innovation, and digital creativity also act as a driver to continually improve their goods and services to satisfy visitors and win their loyalty. Creativity, defined as the process of coming up with innovative and practical ideas, is a key element in the growth of both businesses and people. In this context, digital innovation and creativity are also a necessity for the tourism industry.

Digital Technologies Supporting Creativity in Tourism

The opportunities and challenges of digitizing tourism are aimed at creating digital competencies and capabilities, changing mindsets, being creative and innovative, creating new consumer and producer relations, adopting newer and healthier habits, creating newer and healthier value chains, facilitating the personalization of the tourist experience, raising funds, contributing towards the creation of new tourist destinations, improving infrastructure, and providing political support (Konstantinova, 2019). The dynamic changes and challenges call for the development of new business models, and the increasing competition in the tourism industry calls for a thorough review of the current strategies and practices. The role of digitalization in this process is of key importance. Digital technologies are transforming the tourism business, products and experiences, industry ecologies, and destinations, and as a result, the tourism industry is being drastically transformed (Barykin et al., 2021). This digital transformation involves re-examining business models and processes to improve businesses and provide customers with a new way to interact with businesses (Samuels, 2022). In the tourism industry, different trends of digital transformation are becoming apparent. IoT, AI, VR, big data, analytics, client engagement, social media, robotic apps, blockchain technology, and sharing economy platforms are some examples of these transformations (Imtiaz & Kim, 2019).

Smart applications of the technologies developed by this transformation (e.g., cloud computing, location-based services, virtual reality, augmented reality) in the context of tourism enhance the tourism experience and services while at the same time creating new tourism business models (Abdullayeva, 2019). These business models are fostering creative tourism, which UNESCO defines as participatory learning about the art, heritage and unique character of a place by engaging with its population and living culture. For instance, location-based services, mobile apps, cultural e-trails and augmented reality are revolutionising the way people discover destinations, while social media platforms such as Facebook, Trip Advisor and YouTube are also transforming the travel experience through the seamless sharing of reviews and recommendations (Castro, 2012). The growth and use of advanced information and communication technologies has been a factor in the emergence of creative tourism (ICT). Today, tourists are no longer just consumers and visitors to concrete and intangible cultural attractions, but active co-producers and co-creators of their own experiences (Trinchini & Spyriadis, 2019). Tourism destinations are developing increasingly sophisticated tourism strategies to improve their competitiveness and urban development. Tourism businesses can benefit from IoT in eight distinct ways: personalization, real-time information, flight experience, operational efficiency, customer service, automation, tourist experience, and maintenance (Dave, 2018). Research on business ecosystems shows that organizations seek digitalization in two ways: first, to better understand markets and customers, and second, to use technology to operate in digital ecosystems and along value chains (OECD, 2020).

According to Berman, there are also three basic strategies that constitute digital transformation strategies (Berman, 2012):

- Paying attention to the value proposition to the customer,
- Business model transformation,
- Merging of these approaches through the planning of operations for supply and delivery while simultaneously changing the customer value proposition.

In this way, companies that can optimize both the physical and digital components by developing new business models based on consumer demand will have the first choice of talent, partners, and resources. The tourism business is one of those that have embraced new technologies in an amazing way. Technology is making it possible for tourism organizations to develop new business models. Organizations are able to innovate and stay ahead of the competition by using technology to change the value proposition and business model for their customers. As a result, creativity in tourism may be seen in the four 4Ps: creative person, creative process, creative product, and creative environment (Horng et al., 2015). Creativity in tourism is achieved when the creative process is incorporated into the design of the creative activity, the creative people are involved, and the creative environment is used to create creative tourism products (tourist attractions) for the creative people (Florida, 2012).

Case Studies and Future Trends in the Relationship of Creativity and Tourism

Tourism businesses and destinations are gaining competitive advantage in proportion to their ability to keep pace with the rapid development of information and communication technologies and the spread of the Internet. Technological innovations and the use of these innovations in creative ways have a direct impact on the tourism sector. As tourism businesses and destination management organizations turn to new technologies to use digital efficiently and creatively (Leung et al., 2013), customers seek, find, and share the information needed to select tourism products and destinations, as well as the opportunity to access new technological experiences (Pabel & Prideaux, 2016). Based on the importance and necessity of the use of digital and creativity in the tourism industry, we provide a few examples of technological applications used in the tourism industry.

Kazakhstan's aim is to interact digitally with tourists through game apps. For that purpose, it has launched an online game where tourists can explore the 100 sacred sites of the country. The game is a kind of digital money that can be earned through the game's gameplay. Those who make progress in visiting these 100 points in the game can earn titles according to their progress. The digital money earned in the game can also be converted into real money, allowing tourists, for example, to enjoy their first-class flights. Virtual Reality (VR) applications, which are another technological application, allow tourists to visit places that they do not want to visit for various reasons but which they still want to see, such as Chernobyl. Augmented Reality (AR), another application, allows tourists to visit destinations where a film they're watching or a novel they're reading is set, without actually going there. For instance, the 'Hobbiton AR App' allows you to experience some of the locations in 'The Hobbit' through a digital portal. In addition, augmented reality apps could enhance a customer's experience while visiting a tourist attraction (Stylos & Vassiliadis, 2023; Walsh, 2019).

Food, souvenirs, clothes, eyeglasses, hotels, airplanes, and many other products and structures are being designed and produced in the digital environment for use in the real world, thanks to '3D printers', which are already being used in the tourism sector and are expected to become much more widespread in the future (Beltagui et al., 2021; Ekren et al., 2023). Sushi Singularity, for example, analyzes customers' biometric and genomic data to produce 'new sushi' containing required vitamins. This technology enables physical objects to be created quickly, cheaply, and personalized (Muzdakis, 2020).

The use of 'robots', 'chat bots,' and 'artificial intelligences' is another example of how tourism is going digital. Guests will encounter robots and artificial intelligence in almost all areas of the tourism sector. Currently, some hotels, airports, restaurants, and shopping malls have robots and robotic applications working in different units. Some of these robots greet customers at reception, handle check-in and check-out, provide room keys, and record for face and voice recognition. Some robots even prepare meals for guests, while others help guests figure out where to go and what types of food are available in which restaurants and cafés. These robots are used in shopping malls, airports, and restaurants, as well as hotels (Ivanov & Webster, 2020). The concepts of 'smart tourism' and 'smart hotel' have also emerged with digitalization. Mobile keys, digital check-in, curtains, lighting, room temperature, air conditioning, TV, music, and video playback are quickly done by clicking or using a voice assistant (Acikgoz et al., 2023; Stylos et al., 2021). Slippers that find their own place and the room returning to the previous order when the guest leaves are some of the examples that can be used to illustrate the point reached by digitalization.

As part of their centenary celebrations in 2019, The Hilton Group has released a futuristic report, predicting what the tourism industry will be like in 100 years' time. The report highlights how the hospitality and tourism industry will be in a state of flux, with sustainability, personalized design, food, and human relations being themes that will generally be at the forefront of technology. The prominent technology-driven elements in the report include the following (Hilton, 2019):

- Using microchips placed under the skin, it will be possible to wirelessly control every device in the environment, from heating to lighting, from entertainment to all sorts of other details, whenever and wherever needed.
- Features such as 3D-printed dinners and room service will make for a uniquely personalized experience.
- Each guest's biometric data is provided to the chefs, and meals are automatically prepared based on preferences and nutritional needs.
- Digitized exercises such as racing a virtual sea turtle in a swimming pool or climbing the digitized version of Mount Everest will be highly personalized. Furthermore, the energy from these individual exercises is used to power the hotel, creating an eco-friendly, circular system.
- Technology will provide the necessary and important time in which guests can focus on connecting with each other and creating moments they will never forget.
- The only businesses that will survive in the next century will be those that focus on sustainability and social responsibility.
- As for many years, 100 years from now, hotel guests will seek comfort and relaxation in their hotel stays, and to enhance this experience, hyper-personalization will allow guests to create and relax in environments tailored to their specific needs.

Although the benefits of the use of technology are very important in terms of the tourism sector, technological developments can bring some disadvantages as well as the benefits that it offers (Styvén & Wallström, 2019). The technological developments listed above, some of which are currently being used and some of which are planned to be used in the near future, may have a negative impact on people in physical, mental, spiritual, material, and intellectual terms, in addition to the benefits they offer. Businesses can engage in constant 'message bombardment' of consumers through the benefits of technology. A new market has already been created in tourism by the negative effects of technology. People in need of a break from the highly digitalized world that digital technologies have created can go on digital detox camps (Özdemir & Göktaş, 2021). Besides, no amount of 360-degree video can replace experiences such as the tradition of rose- picking that has existed in Bulgaria for more than a century. Nevertheless, digitalization has been a catalyst for the opening of new horizons in the world of tourism and has been a catalyst for the creation of new and beautiful stories. In this regard, companies that can interpret the possibilities offered by technology from a consumer perspective in the future will have the chance to gain a competitive advantage to the extent that they can combine these interpretations with creativity.

Conclusion

The digital transformation of businesses became more important, and in some cases even inevitable, with the new norms and standards that emerged with the COVID-19 pandemic. Businesses that started to digitize before the epidemic were more resilient during the COVID-19 pandemic and managed to re-emerge stronger afterwards (Kutnjak, 2021). The process in question is this: rapidly changing consumer demands in the tourism and travel sector, differentiation from competing markets, revealing their competitive side, maintaining communication with consumers during the pandemic period, and informing them about health and hygiene measures. While it is more advantageous for countries that rapidly change consumer demands in the tourism and travel sector, it has also increased the need for digitalization without exception (TUROFED, 2020). This means that all sectors need to be digitalized to become competitive. In terms of destinations and businesses, this digital transformation should be designed as a fully integrated digitalization. Simultaneously, training of both employees and other stakeholders in the sector will be necessary to make sure that the digital culture in the destination and business can be correctly and effectively utilized and that these digital habits are routinely practiced (Pranita, 2018).

Research shows that investing in digital assets helps businesses differentiate themselves to their guests. It makes a destination more appealing and predictable for travelers and also fosters creativity and competitiveness (Teng et al., 2022). While digital channels start the travel process before the tourist sets out, digital gives tourism professionals the opportunity to tell the story of your business before the guest ever reaches your business or destination. Furthermore, the insights offered by digital platforms provide a better understanding of travel trends and support business decisions. New technologies are creating important links between individuals as well as between individuals and businesses, increasing the level of interaction and making it easier to connect. Businesses encourage creativity and adaptation to digital processes to support the generation of new ideas and solutions to emerging problems or opportunities that can be evaluated. Due to the increasing interest of people and businesses in the creative process and the shift of the economy towards a new digital era, new initiatives and applications in this field are becoming more and more dynamic (Smailhodžić & Berberović, 2021). Market efforts created through the use of creativity and new technologies can create a competitive advantage for businesses and destinations, as they help businesses and destinations to distinguish themselves in terms of image creation in a largely homogeneous market.

As the world becomes more complex and at the same time more accessible due to technological developments, the business areas and sectors related to tourism are constantly changing and the presence of new technologies forces companies to adapt to this new environment with a creative perspective. In this new digital era, the creativity and ability of tourism operators to use these technologies in their marketing efforts through digital technologies are essential in order to promote their market position and develop competitive edge (Magadán & Rivas, 2018).

According to Magadán and Rivas (2018, p. 73), there are four main themes that highlight the need for creativity in terms of tourism businesses and destinations: (1) globalization—the increase in competition due to globalization and the progressive disappearing of borders in accessing new markets; (2) digitalization—the evolution that affects all sectors and systems from an organizational point of view, especially in the direction of communication, marketing, and information technology; (3) new consumers—the presence of tourists who are more socially, culturally, or environmentally aware, more experienced, and more interested in the context of the destinations and businesses they visit, and more open to new experiences; (4) new trends and fashions—destinations visited by tourists throughout their lives and tourism in these destinations the need to gather different experiences from businesses. Not only post-pandemic but also in normal conditions, the new conditions listed by the basic points make creativity and marketing with new tourism/accommodation technologies essential.

Recommended Reading

Berman, S. J. (2012). Digital transformation: Opportunities to create new business models. *Strategy & Leadership, 40*(2), 16–24.

Das, K., Patel, J. D., Sharma, A., & Shukla, Y. (2023). Creativity in marketing: Examining the intellectual structure using scientometric analysis and topic modeling. *Journal of Business Research, 154*, 113384.

Hoffmann, J., Ivcevic, Z., & Brackett, M. (2016). Creativity in the age of technology: Measuring the digital creativity of millennials. *Creativity Research Journal, 28*(2), 149–153.

McIntyre, R. P. (1993). An approach to fostering creativity in marketing. *Marketing Education Review, 3*(1), 33–36.

Mkwizu, K. H. (2019). *Digital marketing and tourism: Opportunities for Africa. International Hospitality Review.*

Sofronov, B. (2019). The development of marketing in tourism industry. *Annals of Spiru Haret University. Economic Series, 19*(1), 117–127.

Stylos, N. (2023). Intelligence augmentation via human-AI symbiosis: Formulating wise systems for a metasociety. In *Philosophy of artificial intelligence and its place in society* (pp. 301–314). IGI Global.

Stylos, N., & Zwiegelaar, J. (2019). *Big data as a game changer: How does it shape business intelligence within a tourism and hospitality industry context?* (pp. 163–181). Springer Singapore.

Yoga, I. M. S., Korry, N. P. D. P., & Yulianti, N. M. D. R. (2019). Information technology adoption on digital marketing communication channel. *International Journal of Social Sciences and Humanities, 3*(2), 95–104.

Young, J. G. (1985). What is creativity? *The Journal of Creative Behavior.*

References

Abdullayeva, N. (2019). *The role of ICT in tourism globalisation. Economic and social development: Book of proceedings* (pp. 1086–1090).

Acikgoz, F., Perez-Vega, R., Okumus, F., & Stylos, N. (2023). Consumer engagement with AI-powered voice assistants: A behavioral reasoning perspective. *Psychology & Marketing, 40*(11), 2226–2243.

Barykin, S. E., de la Poza, E., Khalid, B., Kapustina, I. V., Kalinina, O. V., & Iqbal, K. M. J. (2021). Tourism industry: Digital transformation. In *Handbook of research on future opportunities for technology management education* (pp. 414–434). IGI Global.

Battro, A. M. (2009). Digital Intelligence: The evolution of a new human capacity. *Scientific Insights into the Evolution of the Universe and of Life Pontifical Academy of Sciences, Acta, 20*, 2009.

Beltagui, A., Sesis, A., & Stylos, N. (2021). A bricolage perspective on democratising innovation: The case of 3D printing in makerspaces. *Technological Forecasting and Social Change, 163*. https://doi.org/10.1016/j.techfore.2020.120453

Berman, S. J. (2012). Digital transformation: Opportunities to create new business models. *Strategy & Leadership, 40*(2), 16–24.

Boughzala, I., Garmaki, M., & Tantan, O. C. (2020, January). Understanding how digital intelligence contributes to digital creativity and digital transformation: A systematic literature review. In *HICSS* (pp. 1–10).

Castro, C. M. D. (2012). *New technology and creative tourism: A case study for the city of Porto*. Doctoral dissertation.

Das, K., Patel, J. D., Sharma, A., & Shukla, Y. (2023). Creativity in marketing: Examining the intellectual structure using scientometric analysis and topic modeling. *Journal of Business Research, 154*, 113384.

Dave, N. (2018). *8 ways in which IoT is shaping the future of travel industry*. Accessed January 24, 2023, from https://www.digitaldoughnut.com/articles/2018/january/ways-in-which-iot-is-shaping-the-future-of-travel

Dollinger, S. J., Burke, P. A., & Gump, N. W. (2007). Creativity and values. *Creativity Research Journal, 19*(2–3), 91–103.

Dora, Y. M., & Saudi, M. H. (2020). The role of digital marketing with creativity and innovation for sustainable Msmes. *PalArch's Journal of Archaeology of Egypt/Egyptology, 17*(10), 2902–2913.

Ekren, B. Y., Stylos, N., Zwiegelaar, J., Turhanlar, E. E., & Kumar, V. (2023). Additive manufacturing integration in E-commerce supply chain network to improve resilience and competitiveness. *Simulation Modelling Practice and Theory, 122*. https://doi.org/10.1016/j.simpat.2022.102676

Florida, R. (2012). The rise of the creative class, revisited. In *The creative class* (pp. 35–62). Basic Books.

Henriksen, D., Mishra, P., & Group, T. (2014). Twisting knobs and connecting things: Rethinking Technology & Creativity in the 21st Century. *TechTrends, 58*, 15–19. https://doi.org/10.1007/S11528-013-0713-6

Hilton. (2019). *Checking into 2119 – The future of hospitality: A report*. https://stories-editor.hilton.com/wp-content/uploads/2019/06/Hilton100-Checking-Into-2119-Report.pdf(02.01.2023)

Hoffmann, J., Ivcevic, Z., & Brackett, M. (2016). Creativity in the age of technology: Measuring the digital creativity of millennials. *Creativity Research Journal, 28*(2), 149–153.

Horng, J.-S., Tsai, C.-Y., Liu, C.-H., & Chung, D. Y.-C. (2015). Measuring employee's creativity: A new theoretical model and empirical study for tourism industry. *Asia Pacific Journal of Tourism Research, 20*, 1353–1373.

Imtiaz, S., & Kim, D. J. (2019). Digital transformation: Development of new business models in the tourism industry. *Culinary Science & Hospitality Research, 25*(4), 91–101.

Ivanov, S., & Webster, C. (2020). Robots in tourism: A research agenda for tourism economics. *Tourism Economics, 26*(7), 1065–1085.

Janse van Rensburg, C., Coetzee, S. A., & Schmulian, A. (2022). Developing digital creativity through authentic assessment. *Assessment & Evaluation in Higher Education, 47*(6), 857–877.

Kienitz, E., Quintin, E., Saggar, M., Bott, N., Royalty, A., Hong, D., Liu, N., Chien, Y., Hawthorne, G., & Reiss, A. (2014). Targeted intervention to increase creative capacity and performance: A randomized controlled pilot study. *Thinking Skills and Creativity, 13*, 57–66. https://doi.org/10.1016/J.TSC.2014.03.002

Konstantinova, S. (2019). Digital transformation in tourism. *KNOWLEDGE-International Journal, 35*(1), 188–193.

Kumar, V. (2018). Transformative marketing: The next 20 years. *Journal of Marketing, 82*(4), 1–12. https://doi.org/10.1509/jm.82.41

Kutnjak, A. (2021). Covid-19 accelerates digital transformation in industries: Challenges, issues, barriers and problems in transformation. *IEEE Access, 9*, 79373–79388. https://doi.org/10.1109/ACCESS.2021.3084801

Lee, M. R., & Chen, T. T. (2015). Digital creativity: Research themes and framework. *Computers in Human Behavior, 42*, 12–19.

Leung, D., Law, R., van Hoof, H., & Buhalis, D. (2013). Social media in tourism and hospitality: A literature review. *Journal of Travel & Tourism Marketing, 30*, 3–22.
Ludwig, A. M. (1989). Reflections on creativity and madness. *American Journal of Psychotherapy, 43*(1), 4–14.
Magadán, M., & Rivas, J. (2018). *Innovation, creativity and tourism in the Emerald handbook of entrepreneurship in tourism, travel and hospitality* (pp. 71–88).
Matejun, M. (2018). The process of opportunities exploration and exploitation in the development of SMES' innovativeness. *Management and Production Engineering Review, 9*, 3–15. https://doi.org/10.24425/119529
McIntyre, R. P. (1993). An approach to fostering creativity in marketing. *Marketing Education Review, 3*(1), 33–36.
Mithas, S., & McFarlan, F. W. (2017). What is digital intelligence? *IT Professional, 19*(4), 3–6. https://doi.org/10.1109/MITP.2017.3051329
Mkwizu, K. H. (2019). Digital marketing and tourism: Opportunities for Africa. *International Hospitality Review.*
Muzdakis, M. (2020). *Tokyo restaurant offers 3D-printed sushi tailored to your health needs.* Accessed January 3, 2023, from https://mymodernmet.com/3d-printed-sushi-singularity/
OECD. (2020). Chapter 2: Preparing tourism businesses for the digital future. In *OECD tourism trends and policies 2020.* Accessed January 24, 2023, from https://www.oecd-ilibrary.org/sites/f528d444-en/index.html?itemId=/content/component/f528d444-en
Onur, D., & Zorlu, T. (2017). Yaratıcılık Kavramı ile İlişkili Kuramsal Yaklaşımlar. *Itobiad: Journal of the Human & Social Science Researches, 6*(3).
Özdemir, M., & Goktas, L. (2021). Research trends on digital detox holidays: a bibliometric analysis, 2012–2020. *Tourism & Management Studies.* https://doi.org/10.18089/tms.2021.170302
Pabel, A., & Prideaux, B. (2016). Social media use in pre-trip planning by tourists visiting a small regional leisure destination. *Journal of Vacation Marketing, 22*(4), 335–348.
Pranita, D. (2018). Digitalization: The way to tourism destination's competitive advantage (case study of Indonesia marine tourism). In *The 2nd ICVHE the 2nd international conference on vocational higher education (ICVHE) 2017 "The importance on advancing vocational education to meet contemporary labor demands"* (Vol. 2018).
Rhodes, M. (1961). An analysis of creativity. *The Phi delta kappan, 42*(7), 305–310.

Samuels, M. (2022). *What is digital transformation? Everything you need to know about how technology is reshaping business*. Accessed January 24, 2023, from https://www.zdnet.com/article/what-is-digital-transformation-everything-you-need-to-know-about-how-technology-is-reshaping/

Smailhodžić, E., & Berberović, D. (2021). Digital creativity: Upgrading creativity in digital business. In M. Soltanifar, M. Hughes, & L. Göcke (Eds.), *Digital entrepreneurship – Impact on business and society*. Springer.

Sofronov, B. (2019). The development of marketing in tourism industry. *Annals of Spiru Haret University. Economic Series, 19*(1), 117–127.

Sousa, M. J., & Rocha, Á. (2019). Strategic knowledge management in the digital age. *JBR*. Special Issue Editorial.

Stylos, N., Fotiadis, A. K., Shin, D. D., & Huan, T. C. T. (2021). Beyond smart systems adoption: Enabling diffusion and assimilation of smartness in hospitality. *International Journal of Hospitality Management, 98*. https://doi.org/10.1016/j.ijhm.2021.103042

Stylos, N., & Vassiliadis, C. A. (2023). Gamification design: Toward developing image perception scales for generation Z consumers. *International Journal of Contemporary Hospitality Management*. https://doi.org/10.1108/IJCHM-12-2022-1616

Stylos, N., & Zwiegelaar, J. (2019). *Big data as a game changer: How does it shape business intelligence within a tourism and hospitality industry context?* (pp. 163–181). Springer.

Stylos, N., Zwiegelaar, J., & Buhalis, D. (2021). Big data empowered agility for dynamic, volatile, and time-sensitive service industries: The case of tourism sector. *International Journal of Contemporary Hospitality Management, 33*(3), 1015–1036.

Styvén, M., & Wallström, Å. (2019). Benefits and barriers for the use of digital channels among small tourism companies. *Scandinavian Journal of Hospitality and Tourism, 19*, 27–46. https://doi.org/10.1080/15022250.2017.1379434

Teng, X., Wu, Z., & Yang, F. (2022). Research on the relationship between digital transformation and performance of SMEs. *Sustainability*. https://doi.org/10.3390/su14106012

Trinchini, L., & Spyriadis, T. (2019). Towards smart creative tourism. In V. Katsoni & M. Segarra-Oña (Eds.), *Smart tourism as a driver for culture and sustainability. Springer proceedings in business and economics*. Springer.

TUROFED. (2020). *Businesses growing with digital*. Accessed December 19, 2022, from https://www.turofed.org.tr/panel/upload_system/pages_file/cee2327efb28b202b1ec90d5deb5d161.pdf

Waller, S. (2020). *What is digital intelligence?* www.simonwaller.com.au; https://www.simonwaller.com.au/digital-intelligence

Walsh, M. (2019). *Be transported into new virtual worlds with AR portals*. Accessed December 22, 2022, from https://www.foundry.com/insights/vr-ar-mr/virtual-worlds-ar-portals

Yahyagil, M. Y. (2001). Örgütsel yaraticilik ve yenilikçilik. *IÜ İsletme Iktisadi Enstitüsü, Yönetim Dergisi, 12*(38), 7–16.

Yoga, I. M. S., Korry, N. P. D. P., & Yulianti, N. M. D. R. (2019). Information technology adoption on digital marketing communication channel. *International Journal of Social Sciences and Humanities, 3*(2), 95–104. https://doi.org/10.29332/ijssh.v3n2.297

Young, J. G. (1985). What is creativity? *The Journal of Creative Behavior*.

7

Online Travel Behavior of Millennials Versus Gen Z: A Comparative Analysis

Georgia Zouni, Ioannis Katsanakis, and Christos Klintsaris

Introduction

Targeting Millennials and Gen Z, two of today's most significant demographic groups, is challenging for tourism destinations and organizations. To reach these markets with integrated tourism experiences and services, they must first identify and comprehend their behavioral patterns and differences. Despite those groups' importance to the travel industry, the tourism sector occasionally seems to ignore these travel behaviors. Thus, this chapter will look into how technology has affected Millennials' and Gen Z's travel behavior. The overall goal of this chapter is to compare the differences in online travel behavior of two of today's most important

G. Zouni (✉) • I. Katsanakis
University of Piraeus, Piraeus, Greece
e-mail: ikatsana@unipi.gr

C. Klintsaris
Hellenic Open University, Patras, Greece
e-mail: std140225@ac.eap.gr

N. Stylos et al. (eds.), *Contemporary Marketing Management for Tourism and Hospitality*, https://doi.org/10.1007/978-3-031-65049-9_7

demographic groups, Millennials and Gen Z. The chapter specifically examines (a) the tourism behavior of Millennials and Generation Z; (b) the impact of technology on their tourism behavior; and (c) the impact of the COVID-19 healthcare crisis on their travel behavior. A primary survey with a self-administered questionnaire was developed and implemented to serve the above purposes and research questions. Evidence from primary research studies reveals intriguing findings and conclusions about the two generations' tourist behavior, the online media they use, the role a hotel brand plays in the selection process, and their travel behavior in relation to the COVID-19 health crisis. The chapter ends with recommendations for tourism destinations and organizations targeting these important demographic groups with integrated tourist experiences and services by knowing these distinctions and similarities.

Generations Y and Z: Millennials and Gen Z

According to Kim et al. (2015), a generation is a group of people who are distinct from other generations due to the attitudes and behaviors they share. Regarding the generational borders of Generation Y, experts are still at odds. Dolot (2018) suggested that Generation Y refers to people born between 1980 and 1995; however, some experts only consider people born between 1977 and 1995 to be part of this generational cohort (Huang & Petrick, 2009). Generation Z's borders are likewise different. Turner (2015) stated that people born between 1993 and 2005 were affected, but other experts stated that people born after 1995 were affected (Dolot, 2018). Although there are various generations designated, these are the most generally accepted ones: Generation Y, or "Millennials" ranges from 1981 to 1995, and Generation Z ranges from 1996 to 2010 (Stylos et al., 2021).

Millennials are individuals who work and earn their own money, account for more than half of the global working population, and spend a significant portion of it on travel and recreation (Amaro et al., 2016). They are expected to outnumber the Baby Boomers as the largest adult population still alive by 2019, reaching 73 million (Pew Research Centre, 2019).

The most prominent event that defines Generation Y is the 9/11 terrorist attacks. However, the Digital Revolution was the catalyst that truly transformed this generation (Raval, 2020). This generation is commonly referred to as tech-savvy or the "connected generation" because they were born and raised in the digital and internet eras and have spent their entire lives utilizing technology (Howe & Strauss, 2003). They are the first high-tech generation, known as "Digital Natives". They are thus active users of social media, with one particularly interesting finding being the need for these pages to load quickly, as Millennials lose interest in social platforms that take more than 4 seconds to load (Newswire, 2019). Easy and direct access to information is a constant requirement for Millennials, who are often described as the generation that demands the answer before even asking the question (Sjekavica et al., 2017).

One out of every four Millennials spends more than 3–5 hours per day on social media, with nearly half spending at least 1–3 hours per day (Schiopu et al., 2016). Furthermore, Millennials like revealing the details of their everyday activities and way of life to their online contacts more than any other generation, earning them the reputation of being an image- and video-obsessed generation. It is therefore easy to explain Millennials' preference for online platforms such as Facebook, Instagram, and YouTube, which have experienced tremendous growth in recent years when Millennials are active on the Internet (Schiopu et al., 2016). According to Team T. (2016), Millennials are more likely to leave a website that contains hazy images or long texts and descriptions that are rarely read. Videos that reflect the business premises are another element that may draw the attention of Millennials, who are a visually-oriented generation that is greatly influenced by the image and quality it brings. Furthermore, due to the generation's preference for easy access to any type of information, call-to-action buttons, such as the one for making a reservation, should be at the top of the page and visible in all aspects, even if the menu or requested information changes.

On the other hand, the members of Generation Z, or "Gen Z," refer to those people who were born between 1995 and 2010. This is also known as the generation of "Real Digital Natives" (Francis & Hoefel, 2018). This new generation was expected to inherit a prosperous economy with minimal unemployment, but this seems to have changed because of COVID-19, which has altered the world's social, political, and economic landscape, reduced options, and caused several phenomena like great resignation. Often referred to as the "True Gen," this generation is driven by causes and the search for the truth. Most of their choices are made in a very rational and practical way, and they strongly believe that dialogue can solve problems and make the world a better place (Francis & Hoefel, 2018). Several members of Gen Z pursue socially conscious employment and launch their own enterprises with an emphasis on social and environmental sustainability. Gen Z is also renowned for its entrepreneurial spirit and desire to have a positive impact on society. Moreover, Gen Z is a highly global and connected generation with diverse perspectives on culture, identity, and social issues (Lifintsev et al., 2019). Above all, Gen Z is regarded as the generation with an innate talent for all things technological because it is the first to have been born in the digital age at the native level (Yu, 2022). In detail, Gen Z is very reliant on the internet and digital technology, with 60% finding it difficult to go more than four hours without it (Ad Tech Daily, 2020). This dependence on the Internet is unquestionably leading all other generations in terms of technology use (Business Wire, 2020). Generation Z, in particular, displays distinct preferences in engaging with smart retail and digital systems, as evidenced by their expectations of seamless interactions in technology-enhanced environments (Priporas et al., 2017). This techsavvy generation values open data, which facilitates decision-making processes through predictive analytics and recommender systems (Pantano et al., 2017, 2019).

Social media use is part of Gen Z's digital habit (Janssen & Carradini, 2021). They are frequently referred to as the "social media generation" and are ardent users of social media sites (Stahl & Literat, 2022). They favor visual platforms like Instagram, Snapchat, and TikTok rather than being as engaged on Facebook as earlier generations (Jain & Arakkal, 2022). Gen Z uses social media for various things, such as keeping in touch with loved ones, finding entertainment and news, and learning about new goods and services (Euajarusphan, 2021; Stylos & Vassiliadis, 2023). Recent research suggests that Generation Z's use of social media and technology may have an impact on their communication preferences and online interactions (Sarkar et al., 2018). They are considered to be quite independent and cooperative, which might translate into how they act online (Fox, 2019). Also, the way they interact with others on social media sites may change as a result of their varied viewpoints on culture, identity, and social concerns (Stanford News, 2022). Overall, their identity and online presence are intrinsically intertwined, demonstrating their strong connections. Some individuals in Gen Z perceive the internet as an extension of who they are and feel emotionally tied to it, according to a study by Turner (2015) on emotional attachment to the Internet. Online dating and relationship retention have a significant impact on people's social lives.

Travel Behavior of Millennials and Gen Z

According to recent studies, the travel behavior of Millennials and Gen Z has significantly evolved over the past few years, affecting the tourism industry (Deloitte, n.d.). Overall, it is notable how the travel habits of Millennials and Gen Z differ from one another. Although they have certain things in common, they use technology very differently and have different objectives.

Regarding travel preferences, Gen Z is more inclined toward urban travel and visiting well-known tourist attractions than Millennials, who often favor adventure and off-the-beaten-path destinations (Walia & Jasrotia, 2021). Millennials favor convenience, customization, and affordability in their travel experiences, according to a study by Expedia (2018). On the other hand, Gen Z travelers have particular traits that set them apart from other age groups when it comes to travel, according to the European Tourism Commission (ETC, 2020). The major traits of Generation Z that have been highlighted by the ETC as being important factors for travel service providers to take into account are these: (a) over time, they are more likely to return to places and contribute value to them; (b) they are curious about visiting new places; (c) they frequently use electronics while traveling; (d) they aim to have a positive impact in the countries they travel to and seek out personal cultural benefits; and (e) travel service providers must consider the distinctive needs of travelers as the travel business develops.

In terms of booking patterns, Generation Z travelers are more receptive to receiving intriguing offers via direct channels like email and messaging (Hotelmize, 2022). They accept the direct offers, think carefully about their worth, and act on them by planning new travels. Millennials, on the other hand, are more inclined to value convenience and cost-effective travel options, such as low-cost airlines and shared lodgings (Starcevic & Konjikušić, 2018). Moreover, Millennials are more inclined to rely on social media for inspiration and advice related to travel, with 72% of Millennials reporting that social media influences their travel decisions (Werenowska & Rzepka, 2020). In addition, Gen Z frequently makes last-minute reservations, whereas Millennials typically plan their vacations months in advance (Baltescu, 2019). In addition, it appears that Gen Z prefers OTAs to all other online booking options for making hotel, flight, and vehicle reservations (Hotelmize, 2022).

While both generations value authentic and distinctive experiences, Gen Z is more likely to seek out local cuisine and cultural encounters than Millennials, who could place a greater emphasis on outdoor recreation and adventure sports (Kamboj & Sharma, 2016). Gen Z is more inclined than Millennials to do it with friends, and they value real experiences above luxurious lodging (Marin-Pantelescu, 2021). Also, they are more inclined to travel abroad and to pick environmentally and socially responsible vacation spots. Regarding how technology affects travel behavior, digital platforms and technologies have increased travel affordability and accessibility, especially for younger generations. Millennials and Gen Z are heavy users of social media and online booking platforms, and they prefer to plan and book their vacations online. As a result, the tourism sector is always evolving to meet the needs of Millennials and Generation Z (Fox, 2020). Moreover, Gen Z tends to be more impulsive travelers (Liu et al., 2022). A recent survey by Hotelmize (2022) identified that Gen Z travelers are more flexible than Millennials when it comes to COVID-19 safety regulations when traveling. Despite the fact that the global epidemic affected travelers and their plans everywhere, Gen Z appeared to be more tolerant of the difficulties the virus brought about. For instance, people in other cohorts favor local destinations, less congested locations, and remote hotels.

Moreover, Gen Z's engagement with city-related social media is linked to their perceptions of city image and brand personality, highlighting the importance of social and economic contexts (Priporas et al., 2020). In contrast, Millennials tend to engage more with AI-powered voice assistants, reflecting their interest in personalized travel experiences driven by advancements in AI and behavioral reasoning (Acikgoz et al., 2023). Both generations are influenced by big data and blockchain technology, which enable real-time service optimization and transparency in the tourism sector (Stylos et al., 2021; Acikgoz et al., 2024), thus enhancing their travel decisions in increasingly dynamic, data-driven environments (Stylos, 2020; Stylos et al., 2021). But above all, the disparities between these generations' travel habits can be largely attributed to their differing use of technology (Chen et al., 2021). Although Gen Z is more likely to use social media sites like Instagram and TikTok to search for destinations and organize their vacations, both rely significantly on technology for travel planning and booking (Perčić & Spasić, 2021). Lu and Stepchenkova (2021) found that whereas Generation Z likes to explore and book trips directly through travel service providers' websites, Millennials are more inclined to explore lodging and activities on travel review websites like Tripadvisor (Ana & Istudor, 2019). Both groups use mobile applications, online reviews, and social networking sites to find information and book travel (Chen & Xie, 2017). In addition, Gen Z favors using social media influencers and peer recommendations when making vacation suggestions (Pop et al., 2022; Priporas et al., 2017, 2020). According to recent studies (Park & Kim, 2020; Kwon & Oh, 2020), social media marketing initiatives have a favorable effect on Generation Z's travel habits. Finally, technological improvements have profoundly impacted both generations' travel behavior. Artificial intelligence, virtual and augmented reality, and mobile applications are all increasingly being used to customize travel experiences and raise client satisfaction (Gretzel & Yoo, 2020). Understanding these generational nuances allows for the development of tailored marketing strategies and technological solutions that meet the unique needs of each group in the digital age (Stylos et al., 2021; Stylos & Vassiliadis, 2023).

Online Travel Behavior Between Millennials and Gen Z: Evidence from Primary Research

A quantitative primary survey with a self-administered questionnaire was conducted to explore the differences in online travel behavior between Millennials and Gen Z. The study aimed to gather information on the travel behavior of Millennials and Gen Z, the effects of technology on this behavior, and the effects of COVID-19, with the following research questions:

(a) What is the travel behavior of these generation groups?
(b) How does technology affect Millennials' and Gen Z's travel behavior?
(c) How has the COVID-19 healthcare crisis affected Millennials' and Gen Z's travel behavior?

The study's target demographic was Gen Z and Millennials who had recently taken leisure trips. The sampling frame was gathered from travel social media sites and other sources that meet Generation Z and Millennial travelers' needs. In sum, 333 questionnaires were answered from the sample, which either belongs to the Millennial generation (date of birth 1980–1994) or Generation Z (date of birth 1995–2010). A convenience sample of 333 people who either belong to the Millennial generation (date of birth 1980–1994) or Generation Z (date of birth 1995–2010) from various regions of Greece was selected for the purposes of the research. The questionnaire was published on the Internet and, in particular, on social media in order to gather the sample without any geographical limitation. Specifically, it was published on online groups of tourism interest and mostly on online groups of special interest in digital marketing. The questionnaire consisted of 33 questions. The questions were created to collect data on Millennials' and Gen Z's travel preferences and how they are affected by technology (digital marketing), the reputation of the accommodation, and the COVID-19 healthcare crisis (crisis management), in order to provide answers to fundamental questions arising from the theory. In order to investigate the questions that arose from the goals, the questionnaire is divided into five (5) sections. The verification or rejection of null research hypotheses about generational group differences was examined through the statistical method of Pearson's Chi-square test via the statistical software package SPSS. The analysis of the reliability coefficient of Cronbach's alpha was found to be 0.729, which is consequently a very satisfactory measurement and sample validity level.

Millennials' and Gen Z's Demographic Profile

At first, the sample units that participated in the research are presented. From Table 7.1, it is observed that out of the 333 people who participated in the research, the biggest percentage consists of Millennials (65.2%), which is expected since Millennials today are between 27 and 41 years old. Generation Z has a satisfactory percentage in the research, namely 34.8%. Moreover, ¾ of the sample are women (76%), while men have a percentage of 24%. Regarding the educational level of the participants in the research, the cumulative sum of the graduates of bachelor's and master's degrees is verging on 82.8%, thus acquiring the two first places in the research sample, with high school graduates following with 13.5%. Below, a comprehensive table is presented with the demographic data of the sample (Table 7.1).

Table 7.1 The demographic profile of the sample per generation (Source: Authors)

Generation	Tourist number	Percentage (%)
Generation Z (1995– 2010)	116	34.83
Millennials (1980–1994)	217	65.17
Total	333	100
Gender		
Men	80	24.0
Women	253	76.0
Total	333	100.0
Degree title		
PhD graduate	5	1.5
Master's degree graduate	131	39.3
Bachelor's degree graduate	145	43.5
High school graduate	46	13.8
Currently student	5	4.2
Total	333	100.0

Millennials' and Gen Z's Travel Behavior Profile

Around 51.4% of Millennials and Generation Z take two to four trips in a year. One out of three people (35.7%) only travels once a year, compared to 10.4% who travel more than four times. Just 2.4% of the sample reported having traveled literally little, if anything, during the entire year. The type of destination that the generations of Millennials and Zers choose to travel to is of particular interest. Fifty-two percent of the total sample responded that they often choose alternative destinations, while 48% chose famous destinations when traveling. A more detailed account of the type of tourist accommodation that Millennials and members of Generation Z choose when traveling follows. The majority of the sample (52.7%) responded that they choose hotels as travel accommodations, while rented accommodations follow with 15.4%. The first place belongs to the hotel as the accommodation of preference not only of the sample as a whole but also of each generation separately. Over half of the Millennials (55.3%) choose a hotel as their accommodation, with rented accommodation following with 18.4% and Airbnb with 16.1%. On the same wavelength, almost half of the total of Generation Z (47.8%) stated the hotel as the most preferred accommodation, with rented accommodation as a close second option (20.9%) and Airbnb as a third option (13.9%). The types of accommodations favored by Millennials and Generation Z were compared using the Pearson chi-square test of independence. The Pearson chi-square value result ($P = 0.360 > 0.05$; chi-square value = 8,487) indicates no significant difference between the two generations' preferred types of accommodations.

Having understood the tendency of both generations regarding their preference in hotel accommodations, it is advisable to mention that the great majority of the sample prefers smaller hotel accommodations, at a percentage that is verging on 75.2%, while the realization of a reservation happens, according to 82.3% of the sample, through the internet. Finally, 60.4% of Millennials and 14.6% of members of Generation Z said they consult digital reviews throughout their trip to decide which nearby destinations and attractions to visit, where to eat, and so on. The opinions of relatives and friends (word-of-mouth) and the advertisements online follow with 25.1% and 10.3%, respectively.

Millennials' and Gen Z's Tourist Digital Profile

As shown in the abovementioned literature review, technology is an integral part of the lives of Millennials and Gen Z. The percentages of use of the Internet during the planning and realization of a trip dominate entirely over all the other media. 62.2% of Millennials and members of Generation Z use their own mobile phones to get information about hotel accommodations, and 33.5% use a laptop. In Table 7.2, a comparison is made between the two generations and the means they use to get information about tourist accommodations. The preference for the mobile phone among both generations is more than apparent since, for both generations, the percentage amounts to over half of the total number of respondents, or, in terms of percentages, 65.3% for Millennials and 56.0% for Gen Z. Besides, the result is reflected in the total response of the sample, which indicates that access to the online channels of a hotel's accommodation through mobile phones is very important (59.9%) and important (23.8%). The test result is $P = 0.360 > 0.05$ (chi-square value = 3,210), indicating that there is no significant difference between the two generations' preferred types of accommodations.

Table 7.2 Technological devices the sample uses to get information for the hotel accommodation per generation (Source: Authors)

Generation Z			Millennials		Total	
Other	3 2.6%	42.9%	4 1.9%	57.1%	7 2.1%	100%
Mobile phone	65 56.0%	31.6%	141 65.3%	68.4%	206 62.0%	100%
Tablet	2 1.7%	28.6%	5 2.3%	71.4%	7 2.1%	100%
Laptop	46 39.7%	41.4%	65 30.6%	58.6%	111 33.3%	100%
Total	116	34.8%	217	65.2%	333	100%

Pearson Chi-squared test	Value	*Df*	Asymp. Sig. (2-sided)	Linear-by-linear association
	3,210	3	0.360	0.167

$P = 0.360 > 0.05$ Accepted HH_{00}

The most striking finding is not, however, the extensive use of the mobile phone for getting information, but the reduction of the percentage of mobile phones to 43.5% over laptops (53.5%) when making a reservation. This practically means that both Millennials and Gen Z search for information about hotel accommodations through their mobile phones, but they make the reservation mainly through their laptops.

In an attempt to deeply analyze the influence that is being exerted on Millennials and Gen Z by the online presence of a hotel accommodation, the results showed that online reviews influence to a great extent both Millennials (57.87% of Millennials responded that they are influenced very much) and Gen Z when making a reservation (46.2% of Gen Z responded that they are influenced very much). The official webpage of the hotel constitutes another basic influential factor for both Millennials (45.37% of Millennials responded that they are influenced a lot) and Gen Z (46.49% of Gen Z responded that they are influenced a lot). The order of appearance of hotel accommodations in search engines (SEO) is shown not to influence either positively or negatively Millennials (43.66%) and Gen Z (36.61%).

Another influential means for making a hotel reservation is social media, which influences a lot of Millennials (30.70% of Millennials responded that they get influenced a lot). However, social media seems to influence more Gen Z, since their percentage differs from that of Millennials by 10 percentage points (41.82% of Generation Z participants responded that they get influenced a lot). At the question that was asked regarding whether the sample looks at the social media accounts that the hotel accommodation has before making a reservation, a cumulative percentage of 83.4% responded "Yes" and "Sometimes," while only 16.5% responded that they do not look at the social media accounts in question.

Moreover, it is of utmost interest to discover which social media platforms Millennials and members of Generation Z use more frequently. The Millennials' generation seems to use primarily Facebook, with 63.7%, while the most popular social media platform for Generation Z is Instagram, with 56.5%. The second option for each generation is Instagram for Millennials (36.3%) and Facebook for Gen Z (39.1%). It is worth mentioning that the Pearson Chi-Squared Test result is P =

Table 7.3 Preferred social media platforms of the sample per generation (Source: Authors)

Generation Z			Millennials		Total	
Facebook	18 39.1%	23.7%	58 63.7%	76.3%	76 55.5%	100%
Instagram	26 56.5%	48.0%	33 36.3%	52.0%	59 43.1%	100%
Other	2 4.3%	100%	0 0.0%	0.0%	0 1.5%	100%
Total	47	33.6%	92	66.4%	139	100%

Pearson Chi-squared test	Value	*Df*	Asymp. Sig. (2-sided)	Linear-by-linear association
	10,203	2	0.006	0.001

$P = 0.006 < 0.05$ Rejected HH_{00}

0.0006 < 0.05 (chi-square value = 10,203), indicating that there is a significant difference between the two generations' preferred social media platforms. Table 7.3 summarizes this data:

Another interesting finding regarding social media is that 82% of the total sample follows one to four pages of tourist interest that have social media accounts, and 63.3% of the total sample follows one to four travel bloggers.

Finally, 74.9% of the sample as a whole said they were more likely to believe social media posts made by users that are related to the hotel than they would believe hotel advertisements. As regards the factors that influence their reservation preference for a hotel accommodation, the majority of both groups (77.1% of Millennials and 78% of members of Generation Z) responded that they have shown interest in learning more about hotel accommodations after seeing a relevant post from online friends. However, as evidenced by their higher percentage (29.77%) at the reservation stage, Millennials appear to be unaffected by posts, either favorably or unfavorably, whereas Gen Z, at their higher percentage (31.25%), is heavily influenced by friends' posts. Finally, the highest percentage of Millennials and members of Generation Z say that they are not interested in getting information about hotel accommodations after seeing relevant posts from celebrities. Regarding the realization of a reservation, Millennials do not get influenced by celebrities who have visited

Table 7.4 Pearson Chi-squared test results for two generations' influence sources (Source: Authors)

Pearson Chi-squared test	Value	*df*	Asymp. Sig. (2-sided)	Linear-by-linear association	$P > 0.05$
Its official website	1.910	4	0.752	0.746	Accepted HH_0
Its social media	4.400	4	0.355	0.09	Accepted HH_0
Online reviews	6.368	4	0.173	0.173	Accepted HH_0
The order of its appearance at the results of search engines	4.461	4	0.347	0.087	Accepted HH_0
It's a preference from my online friends	3.597	4	0.463	0.099	Accepted HH_0
Its preference from celebrities	24.179	4	0	0	Rejected HH_0

a hotel in order to make a reservation (60.28% of Millennials responded that they do not get influenced at all), while Gen Z, at their higher percentage (34.82%), do not get influenced either positively or negatively. The Pearson Chi-Squared Test results for these aspects (Table 7.4) revealed that there is no significant difference between the two generations' influence sources, except for their influence by celebrities.

Continuing on with the exploration of the digital profile of tourists that belong to the generations of Millennials and Zers, and since it is already concluded that the most popular digital means by which these generations are influenced to make a reservation at a hotel accommodation, it is worth mentioning that, as a whole, after they have decided to make a reservation, they make it through a cooperative webpage of the hotel (67.3%), while 32.7% make the reservation through the webpage of the hotel.

Closing the section about the digital profiles of Millennials and Gen Z, a brief account will be made of the always-increasing travel and hotel applications that lead to the automation of tourism. Specifically, 59% of Millennials and members of Generation Z responded that they have a particular application on their mobile phone when traveling, with the most popular ones being Booking and Tripadvisor. Other applications that were mentioned are Airbnb, Skyscanner, Expedia, and specific airline apps (e.g., Aegean, Ryanair, etc.). It is of particular interest that a large percentage of the sample (87.7%) represents the development of a hotel application, through which most of the actions will be realized (such as check-in or a reservation at a restaurant). Only 12.3% responded that the hotel depends on human contact, and for this reason, it stands against automation.

Millennials' and Gen Z's Tourist Digital Behavior Concerning Hotel Brand

To better understand the factors influencing Millennials and Gen Z to make a hotel reservation, the brand and its ability to attract this particular target group will be examined. On the question of whether the brand of a hotel influences Millennials and Gen Z to make a reservation there, a higher percentage (37%) responded that it influences them a lot. However, a significant portion (34.9%) replied that they are neither positively nor negatively influenced by a strong brand. 71.4% responded that the existence of a strong brand in the hotel industry creates a sense of security for potential clients, with 28.6% having the opposite opinion. More interesting, though, is to find out which online channels are the most efficient for creating, for Millennials and Gen Z, the perception of a strong brand for the hotel accommodation in order to experience the stated security and finally make the reservation. As it arises from the analysis of results per generation, presented in Table 7.5, and in this case of the brand, the two strongest online allies of the hotel accommodation are its official website and social media. In more detail, 43.93% of Millennials and 41.23% of members of Generation Z responded that the official website of the hotel influences them a lot regarding their perception of the brand. Moreover, 33.02% of Millennials and 39.82% of members of Generation Z responded that the content of social media accounts associated with the hotel's accommodations has influenced their perceptions of the brand equally. Furthermore, ICM is also classified in first place among the responses that the two generations gave regarding the brand. The constant message that hotels advertise on all online channels has a significant impact on 30.84% of Millennials and 42.11% of members of

Table 7.5 Pearson Chi-squared test results for two generations' behavior concerning a hotel brand (Source: Authors)

Pearson Chi-squared test	Value	*Df*	Asymp. Sig. (2-sided)	Linear-by-linear association	$P > 0.05$
Its official website	2.011	4	0.734	0.341	Accepted HH_0
Its social media	3.602	4	0.462	0.083	Accepted HH_0
Public relations (media, participation in tourism exhibitions,	3.002	4	0.558	0.229	Accepted HH_0
Its preference by celebrities and influencers	13.135	4	0.011	0.002	Rejected HH_0
The constant message in all its social media (IMC)	7.482	4	0.112	0.463	Accepted HH_0

Generation Z (a difference of almost 12 percentage points). The opinions of the generations are ambiguous regarding the public relations of the hotel and whether these relations ultimately lead to a positive perception of the hotel brand among Millennials and Gen Z. 35.38% of Millennials responded that they are influenced neither positively nor negatively, by the public relations the hotel maintains, while the remainder answered that they get influenced a lot. Both Millennials and members of Generation Z seem not to get influenced to a great extent by celebrities' and influencers' posts, while 36.28% of Gen Z responded that they do not get influenced either positively or negatively. The Pearson Chi-Squared Test results concerning hotel brands (Table 7.5) revealed that there is no significant difference between the two generations' reactions toward a hotel brand and its attributes, except for a hotel's brand preference by celebrities.

Generations' Behavior Concerning COVID-19

Regarding the channel the two generations prefer to get updated on tourism and travel during the pandemic, the striking majority of Millennials and members of Generation Z responded through the Internet (94.6%).

Moreover, concerning the type of accommodation Millennials and Gen Z preferred during the pandemic, both 44.4% of Millennials and 46.1% of Gen Z responded that hotel establishments inspire more security in the era of COVID-19. Finally, the Pearson Chi- Squared test result (P = 0.360 > 0.05, chi-square value = 5.681) confirmed that there is no significant difference between the two generations' behavior concerning the COVID-19 pandemic and their travel behavior.

Conclusion

This chapter offers an in-depth and detailed comparative analysis of Millennials' and Gen Z's online travel behavior, including their tourist behavior, the digital media that influences them, the role of hotel accommodations in their selection process, and the impact of COVID-19 on tourism. According to the study, both generations are heavily involved in tourism, with alternative destinations increasingly appealing over traditional ones. While Gen Z emphasizes cultural immersion and local experiences, Millennials are more likely to seek out adventure activities and outdoor experiences. In terms of travel behavior, Millennials and Gen Z are keen to visit unexplored places where they can have unique experiences to share with their connections online. Also, both generations are concerned about their carbon footprint and prefer environmentally friendly lodging options.

Hotels remain a popular choice for both age groups, with the majority of respondents preferring to stay in hotels while traveling. Apart from hotels, a growing trend is to stay in private accommodations such as Airbnb and HomeAway. These options are especially popular with Millennials looking for a more personalized and authentic travel experience. Furthermore, respondents stated that they trust the opinions of other travelers more than any other form of advertising or marketing, with online reviews having the greatest influence on their decisions to make reservations. Before making a booking decision, Millennials and Generation Z travelers are interested in seeing real-life photos and videos of the hotel. Hotels can capitalize on this by posting user-generated content on their social media accounts and websites.

Regarding digital media, the hotel's official website is the second-ranked internet channel, with a significant influence on Gen Z and Millennials. But social media sites like Facebook and Instagram are also quite powerful, with many respondents saying they use them to plan and conduct research for their travels. Therefore, it is essential for hotels to have a strong social media presence with visually appealing and interesting content that can persuade tourists to book their accommodations.

The COVID-19 pandemic has undoubtedly impacted the tourism habits of all generations, with many travelers opting to postpone or cancel their trips. As a result, hotels must ensure that their websites are mobile-friendly, with simple menus and specialized content on safety protocols and hygiene precautions. As both Millennials and Gen Z say they are heavily influenced by them, businesses should place a high priority on obtaining positive online reviews from customers who visited a tourist attraction during the pandemic. Hotels should consider offering flexible cancelation policies and other incentives to get visitors to make their reservations.

Ultimately, the travel sector is facing significant challenges due to the impact of the pandemic and other worldwide crises and issues, and businesses must change swiftly to satisfy their customers' shifting needs and preferences. There is a great desire for individualized, genuine, and environmentally friendly travel experiences that meet the interests and tastes of Millennials and Gen Z. By staying on top of these trends, the travel industry can thrive and grow in the post-COVID-19 era.

The results of this study have important theoretical ramifications for travel destinations and businesses aiming to attract Millennials and Gen Z with integrated travel experiences and services. The study emphasizes the significance of comprehending the differences in travel preferences and behaviors between these two age groups and the necessity of creating custom travel experiences that cater to their particular interests and values. The study emphasizes the need for hotels and other lodging providers to adapt to Millennial and Generation Z travelers' shifting preferences, who are increasingly looking for individualized and genuine travel experiences. Offering flexible cancelation policies, eco-friendly lodging options, and user-generated content that highlights the distinctive experiences on offer are a few examples of how to achieve this. It might also entail creating effective online marketing and social media strategies that can connect with and engage these digitally savvy travelers.

The practical ramifications of this study's findings suggest, in conclusion, that travel destinations and tourism businesses must adapt quickly to the shifting preferences and values of Millennial and Generation Z tourists. The study emphasizes how crucial it is to address the issues brought on by the COVID-19 pandemic, which has had a profound effect on the travel sector. Hotels and other tourism-related businesses must prioritize safety and hygiene procedures and provide incentives like flexible cancelation policies and favorable online reviews in order to cater to the evolving needs and preferences of Millennial and Generation Z travelers. The travel industry can flourish in the post-COVID-19 era and beyond by providing tailored, genuine, and eco-friendly travel experiences that cater to travelers' interests and values.

Recommended Further Reading

Deloitte. (n.d.). *Gen Z and millennial travel: The wanderlust generations.* https://www2.deloitte.com/us/en/pages/technology-media-and-telecommunications/articles/gen-Z-and-millennial-travel-the-wanderlust-generations.html

Fan, A., Shin, H. W., Shi, J., & (Jade), & Wu, L. (2022). Young people share, but do so differently: An empirical comparison of peer-to-peer accommodation consumption between millennials and generation Z. *Cornell Hospitality Quarterly, 0*(0). https://doi.org/10.1177/19389655221119463

HotelMize. (2022, August 22). Gen Z travelers: How their travel behavior compares to millennials. *HotelMize Blog*. https://www.hotelmize.com/blog/gen-z-travelers-how-their-travel-behavior-compares-to-millennials/

Liu, J., Wang, C., Zhang, T., & Qiao, H. (2022). Delineating the effects of social media marketing activities on generation Z travel behaviors. *Journal of Travel Research*, 00472875221106394.

Prayag, G., Aquino, R. S., Hall, C. M., Chen, N., & Fieger, P. (2022). Is Gen Z really that different? Environmental attitudes, travel behaviors and sustainability practices of international tourists to Canterbury, New Zealand. *Journal of Sustainable Tourism*, 1–22.

Slivar, I., Aleric, D., & Dolenec, S. (2019). Leisure travel behavior of generation Y & Z at the destination and post-purchase. *E-Journal of Tourism, 6*(2), 147–159.
Stylos, N. (2023). Intelligence augmentation via human-AI symbiosis: Formulating wise systems for a metasociety. In *Philosophy of artificial intelligence and its place in society* (pp. 301–314). IGI Global.
Stylos, N., & Zwiegelaar, J. (2019). *Big data as a game changer: How does it shape business intelligence within a tourism and hospitality industry context?* (pp. 163–181). Springer.
TravelPerk. (2022, June 17). *Gen Z travel statistics & trends [Blog post].* https://www.travelperk.com/blog/gen-z-travel-statistics-trends/
Trekksoft. (2022, August 1). *Gen Z: How will the newest generation of travelers impact the tourism industry?* https://www.trekksoft.com/en/blog/gen-z-how-will-the-newest-generation-of-travelers-impact-the-tourism-industry
Uysal, D. (2022). Gen-Z's consumption behaviors in post-pandemic tourism sector. *Journal of Tourism Leisure and Hospitality, 4*(1), 7–7.

References

Acikgoz, F., Perez-Vega, R., Okumus, F., & Stylos, N. (2023). Consumer engagement with AI-powered voice assistants: A behavioral reasoning perspective. *Psychology & Marketing, 40*(11), 2226–2243.
Acikgoz, F., Stylos, N., & Lythreatis, S. (2024). Identifying capabilities and constraints in utilizing blockchain technology in hospitality and tourism. *International Journal of Contemporary Hospitality Management.* https://doi.org/10.1108/IJCHM-07-2023-1083
Ad Tech Daily. (2020). *Generation influence: Gen Z study reveals a new digital reality.* https://adtechdaily.com/2020/07/21/generation-influence-gen-z-study-reveals-a-new-digital-reality/
Amaro, S., Duarte, P., & Henriques, C. (2016). Travelers' use of social media: A clustering approach. *Annals of Tourism Research, 59*, 1–15.
Ana, M. I., & Istudor, L. G. (2019). The role of social media and user-generated-content in millennials' travel behavior. *Management Dynamics in the Knowledge Economy, 7*(1), 87–104.
Baltescu, C. A. (2019). Elements of tourism consumer behavior of generation Z. *Bulletin of the Transilvania University of Brasov. Series V: Economic Sciences*, 63–68.

Business Wire. (2020). *Generation influence: Gen Z study reveals a new digital paradigm*. https://www.businesswire.com/news/home/20200706005543/en/Generation-Influence-Gen-Z-Study-Reveals-a-New-Digital-Paradigm

Chen, X., Li, T., & Yuan, Q. (2021). Impacts of built environment on travel behaviors of Generation Z: A longitudinal perspective. *Transportation*, 1–30.

Chen, C. C., & Xie, K. L. (2017). Exploring online review platforms for travel planning: The impact of generation cohorts. *Journal of Travel Research, 56*(1), 98–112.

Deloitte. (n.d.). *Gen Z and millennial travel: The wanderlust generations*. https://www2.deloitte.com/us/en/pages/technology-media-and-telecommunications/articles/gen-Z-and-millennial-travel-the-wanderlust-generations.html

Dolot, A. (2018). The characteristics of generation Z. *e-mentor, 2*(74), 44–50. https://doi.org/10.15219/em74.1351

Euajarusphan, A. (2021). Online social media usage behavior, attitude, satisfaction, and online social media literacy of generation X, generation Y, and generation Z. *PSAKU International Journal of Interdisciplinary Research, 10*(2).

European Travel Commission. (2020). *Study on Generation Z Travellers*, Accessed on January 5, 2024, via https://etc-corporate.org/reports/study-on-generation-z-travellers

Expedia Group Media Solutions. (2018). *Generations on the move*. https://www.expedia.com/stories/wp-content/uploads/2018/01/Expedia-Generations-on-the-Move.pdf

Fox, E. M. (2019). Mobile technology. *The International Review of Research in Open and Distributed Learning, 20*(2).

Fox, L. (2020, February 2). How millennials and generation Z are shaping the future of travel. *Forbes*. https://www.forbes.com/sites/adamkuylenstierna/2020/02/02/how-millennials-and-generation-z-are-shaping-the-future-of-travel/#1518d25957a9

Francis, T., & Hoefel, F. (2018). True Gen': Generation Z and its implications for companies. *McKinsey & Company, 12*(2), 1–10.

Gretzel, U., & Yoo, K. H. (2020). Use of artificial intelligence in tourism. *Journal of Travel Research, 59*(3), 437–451.

HotelMize. (2022, August 22). Gen Z travelers: How their travel behavior compares to millennials. *HotelMize Blog*. https://www.hotelmize.com/blog/gen-z-travelers-how-their-travel-behavior-compares-to-millennials/

Howe, N., & Strauss, W. (2003). *Millennials go to college*. American Association of College Registrars and Admissions Officers and LifeCourse Associates.

Huang, Y. C., & Petrick, J. F. (2009). Generation Y's travel behaviours: a comparison with baby boomers and Generation X. In *Tourism and generation Y* (pp. 27–37). Wallingford UK: CABI.

Jain, P., & Arakkal, J. (2022). Perceived popularity of TikTok among Gen-Z. *Cardiometry, 24*, 896–902.

Janssen, D., & Carradini, S. (2021). Generation Z workplace communication habits and expectations. *IEEE Transactions on Professional Communication, 64*(2), 137–153.

Kamboj, P., & Sharma, R. (2016). Tourism for tomorrow: Travel trends across generations: From baby boomers to millennials. *Amity Research Journal of Tourism, Aviation and Hospitality, 1*(2), 70–83.

Kim, H., Xiang, Z., & Fesenmaier, D. R. (2015). Use of the internet for trip planning: A generational analysis. *Journal of Travel & Tourism Marketing, 32*(3), 276–289.

Kwon, K. N., & Oh, H. (2020). The influence of social media on travel behavior and destination choice: Generation Z travelers. *Sustainability, 12*(11), 4462. https://doi.org/10.3390/su12114462

Lifintsev, D., Fleşeriu, C., & Wellbrock, W. (2019). A study of the attitude of generation Z to cross-cultural interaction in business. *Information & Media, 86*, 41–55.

Liu, J., Wang, C., Zhang, T., & Qiao, H. (2022). Delineating the effects of social media marketing activities on generation Z travel behaviors. *Journal of Travel Research*, 00472875221106394.

Lu, L., & Stepchenkova, S. (2021). Understanding the travel behavior of millennials and generation Z: A case of North America. *Journal of Travel Research, 60*(4), 874–889.

Marin-Pantelescu, A. (2021). The young travellers behavior facing the Covid-19 pandemic. *Annales Universitatis Apulensis: Series Oeconomica, 23*(1), 17–24.

Newswire. Retrieved February 2, 2019., from http://ezproxy.rit.edu/login?url=https://searchproquest-com.ezproxy.rit.edu/docview/1468403637?accountid=108

Pantano, E., Priporas, C. V., & Stylos, N. (2017). 'You will like it!' using open data to predict tourists' response to a tourist attraction. *Tourism Management, 60*, 430–438.

Pantano, E., Priporas, C. V., Stylos, N., & Dennis, C. (2019). Facilitating tourists' decision making through open data analyses: A novel recommender system. *Tourism Management Perspectives, 31*, 323–331.

Park, S. Y., & Kim, Y. K. (2020). How social media influencers affect destination image and travel intention: A study of young Chinese travelers. *Journal of Travel & Tourism Marketing, 37*(9), 1073–1087. https://doi.org/10.1080/10548408.2020.1813948

Perčić, K., & Spasić, N. (2021). How millennials and generation z organise travel during the COVID-19 pandemic. *Менаџмент у хотелијерству и туризму, 9*(2), 79–94.

Pew Research Center. (2019, February 14). *Millennial life: How young adulthood today compares with prior generations.* https://www.pewresearch.org/social-trends/2019/02/14/millennial-life-how-young-adulthood-today-compares-with-prior-generations-2/

Pop, R. A., Săplăcan, Z., Dabija, D. C., & Alt, M. A. (2022). The impact of social media influencers on travel decisions: The role of trust in consumer decision journey. *Current Issues in Tourism, 25*(5), 823–843.

Priporas, C. V., Stylos, N., & Fotiadis, A. K. (2017). Generation Z consumers' expectations of interactions in smart retailing: A future agenda. *Computers in Human Behavior, 77*, 374–381.

Priporas, C. V., Stylos, N., & Kamenidou, I. E. (2020). City image, city brand personality and generation Z residents' life satisfaction under economic crisis: Predictors of city-related social media engagement. *Journal of Business Research, 119*, 453–463.

Raval, N. A. (2020). *Platform-Living: Theorizing life, work, and ethical living after the gig economy*. University of California, Irvine.

Sarkar, U., Le, G. M., Lyles, C. R., Ramo, D., Linos, E., & Bibbins-Domingo, K. (2018). Using social media to target cancer prevention in young adults. *Journal of Medical Internet Research, 20*(6), e203.

Schiopu, A. F., Padurean, A. M., Tala, M. L., & Nica, A. (2016). The influence of new technologies on tourism consumption behavior of the millennials. *Amfiteatru Economic, 18*(10), 829–846.

Sjekavica, T., Žitnik, M., & Miličević, M. (2017). A novelty model of online accommodation presentation and discovery. *Ekonomska Misao i Praksa, 26*(1), 363.

Stahl, C. C., & Literat, I. (2022). # GenZ on TikTok: The collective online self-portrait of the social media generation. *Journal of Youth Studies*, 1–22.

Stanford News. (2022, January 3). *Know Gen Z.* https://news.stanford.edu/2022/01/03/know-gen-z/

Starcevic, S., & Konjikušić, S. (2018). Why millennials as digital travelers transformed marketing strategy in tourism industry. In S. Starcevic & S. Konjikusic (Eds.), *Why millennials as digital travelers transformed marketing strategy in tourism industry, international thematic monograph tourism in function of*

development of the Republic of Serbia-tourism in the era of digital transformation (pp. 221–224). University of Kragujevac.
Stylos, N. (2020). Technological evolution and tourist decision-making: A perspective article. *Tourism Review, 75*(1), 273–278.
Stylos, N., Fotiadis, A. K., Shin, D. D., & Huan, T. C. T. (2021). Beyond smart systems adoption: Enabling diffusion and assimilation of smartness in hospitality. *International Journal of Hospitality Management, 98*. https://doi.org/10.1016/j.ijhm.2021.103042
Stylos, N., Rahimi, R., Okumus, B., & Williams, S. (2021). *Generation Z marketing and management in tourism and hospitality*. Springer International Publishing.
Stylos, N., & Vassiliadis, C. A. (2023). Gamification design: Toward developing image perception scales for generation Z consumers. *International Journal of Contemporary Hospitality Management*. https://doi.org/10.1108/IJCHM-12-2022-1616
Stylos, N., Zwiegelaar, J., & Buhalis, D. (2021). Big data empowered agility for dynamic, volatile, and time-sensitive service industries: The case of tourism sector. *International Journal of Contemporary Hospitality Management, 33*(3), 1015–1036.
Team, T. (2016). *5 ways to master millennials' booking trends*. Retrieved February 15, 2019, from https://www.triptease.com/blog/ways-master-millennials-booking-trends/
Turner, A. J. (2015). *Generational cohort experiences with training and development: A phenomenological study*. Northcentral University.
Walia, S. K., & Jasrotia, A. (Eds.). (2021). *Millennials, spirituality and tourism*. Routledge.
Werenowska, A., & Rzepka, M. (2020). The role of social media in generation Y travel decision- making process (case study in Poland). *Information, 11*(8), 396.
Yu, M. (2022, January). The acceptance of Chinese generation Z to UGG fur sandals advertisements. In *2021 international conference on social development and media communication (SDMC 2021)* (pp. 643–647). Atlantis Press.

8

The Age of Tomorrow: Exploring Risk-Taking Styles in Travel Across Generations

Eleonora Pantano, Daniele Scarpi, and Mengyun Hu

Introduction

The COVID-19 pandemic outbreak has impacted the tourism industry at the global level for years. The UNWTO (2020) reported that COVID-19 led to a fall of 180 million fewer international tourist arrivals and a US $195 billion loss in export revenues from international tourism between January and April 2020. In 2022, UNWTO (2022) revealed that international tourism remains far below pre-pandemic levels. The

E. Pantano (✉)
University of Bristol, Bristol, UK
e-mail: e.pantano@bristol.ac.uk

D. Scarpi
University of Bologna, Bologna, Italy
e-mail: daniele.scarpi@unibo.it

M. Hu
Newcastle University, Newcastle, United Kingdom
e-mail: mia.hu@newcastle.ac.uk

N. Stylos et al. (eds.), *Contemporary Marketing Management for Tourism and Hospitality*, https://doi.org/10.1007/978-3-031-65049-9_8

reason is that health risks are likely to remain in many travellers' minds, leading to a persistent unwillingness to travel (Duro et al., 2021). Indeed, scholars in the tourism discipline agree that tourism decision-making in risky contexts remains important research (Kim et al., 2022; Sigala, 2020). For instance, Matiza (2020) highlighted that travellers' behaviour associated with the COVID-19 pandemic crisis is an urgent discourse within contemporary tourism research. Along these lines, an in-depth understanding of travellers' risk perceptions and risk-taking behaviour is essential for the tourism industry to adjust its market strategies to recover the tourism economy in the post-COVID-19 era.

This chapter evaluates travellers' behaviour regarding health risks when planning vacations (e.g. COVID-19) by comparing generational cohorts. This chapter first reviews the previous study on risk perception and travel behaviour in the next section. Then, this study critically discusses each generational cohort and their travel behaviour regarding health risks, including Baby Boomers, Generation X, Y, and Z. Next, we discuss risk perception's influence on decision-making and destination image, comparing the generational cohorts. The final section draws conclusions and suggests a future agenda for the tourism industry.

Risk Perceptions and Travel Behaviour

Risk Perception

Risk has been defined as the likelihood or probability of an event occurring and the potential negative consequences associated with that event (Priest, 1990). Three kinds of risks can be identified from the previous literature (Haddock, 1993; Reisinger & Mavondo, 2005), including absolute risk, real risk, and perceived risk. In particular, absolute risk refers to the actual likelihood of an event occurring and is typically measured by business suppliers that undertake safety processes to reduce the actual risk (Im et al., 2021). Real risk refers to the actual risk of something happening, considering all relevant information (Williams & Baláž, 2014). Individuals assess perceived risk in a specific context (Haddock,

1993). It refers to individuals' fear or concern about the potential negative consequences of a particular action or behaviour (Williams & Baláž, 2014). Most tourism and hospitality research concentrates on perceived risk/risk perceptions (Reisinger & Mavondo, 2005; Neuburger & Egger, 2021) because it was an important determinant of customer attitudes and behaviour in tourism (Bae & Chang, 2021; Yang et al., 2017; Ren et al., 2022).

According to Chua et al. (2021), risk perception in tourism refers to how individuals perceive and evaluate the potential risks associated with travelling to a particular destination. It consists of two dimensions: perceived susceptibility and severity (Floyd et al., 2000). Perceived susceptibility refers to travellers' sense of vulnerability to a disease, while severity refers to travellers' perceived seriousness of the disease (Brewer & Fazekas, 2007). Tourism research has confirmed that perceived risk can be an intervening factor that affects travellers' attitudes and behaviour towards the disease (Zheng et al., 2021). The higher the perceived susceptibility and severity of a disease, the more likely the travellers are to take action in corresponding protection, which significantly changes their travel behaviour (Wang et al., 2019).

In recent years, health risk has attracted increasing attention from tourism scholars and practitioners due to the COVID-19 pandemic (Yu et al., 2022; Bae & Chang, 2021). Recent studies have demonstrated that the COVID-19 pandemic has significantly impacted various tourism sectors, reshaping the entire industry (Foo et al., 2021; Hoque et al., 2020). The World Tourism Organization stated that tourism is one of the industries hit the hardest by the pandemic (Dolnicar & Zare, 2020). Considering the relevance and influence of risk on tourists and the industry, it is necessary to understand travellers' behaviour in the era of post-COVID-19.

The Impact of Risk Perception on Travel Behaviour

In tourism, the influences of risks on travel behaviours have been widely discussed for decades (Bae & Chang, 2021). Several studies have explored travellers' perceived risk as an essential determinant of travel intentions and avoidance of potentially risky places (e.g. Neuburger & Egger, 2021; Abraham et al., 2020; Matiza, 2020). Common findings from prior studies revealed a negative association between travellers' perceived risk and travel intentions. It was discovered that higher degrees of perceived risk correlated with lower satisfaction levels, attitude towards a tourist destination, and travel intentions (e.g. Sohn et al., 2016; Neuburger & Egger, 2021; Abraham et al., 2020). Previous studies found that tourists tend to avoid risk and have an individual tolerance threshold to risk (Wang et al., 2019) that, once reached, makes them change their behaviour to reduce the perceived risk (Wang et al., 2019). Travellers can take different actions to mitigate the risks (Law, 2006), such as deferring a holiday, avoiding a destination (Sonmez & Graefe, 1998), and switching to another place (Williams & Baláž, 2014).

Travellers confronted high risks and no proper medical treatment during the pandemic outbreak. Their perceptions of COVID-related risks might cause long-term anxiety about travelling, significantly affecting their travel behaviour (Zheng et al., 2021). Thus, tourism scholars dedicated significant efforts to examining the effect of the pandemic on travellers' behaviour changes in recent years (e.g. Ren et al., 2022; Neuburger & Egger, 2021; Abraham et al., 2020). For instance, Li et al. (2020) indicated that travellers tend to be collectivistic when perceiving high health risks, making them choose domestic rather than overseas travel. Similarly, Zenker & Kock (2020) found that travellers are more likely to choose less populated and rural areas for travel instead of the overcrowded and popular tourist ones. This finding emerges because the rapid spread of coronavirus makes travellers more aware of and evades crowdedness.

Furthermore, Toubes et al. (2021) have found that many travellers have shifted towards more flexible travel options, such as booking refundable or changeable tickets. This shift has also led to an increase in last-minute bookings, as people are more hesitant to plan their trips during the pandemic. Due to the large size and scope of COVID-19, it is important to explore the individuals' risk perceptions and travel behaviour, which play an essential role in reviving the tourism industry after the outbreak (Ren et al., 2022). Moreover, it is important to note that perceived risk is not a static concept; it can vary depending on several factors, such as individual beliefs, values, and age. The next section will discuss travel behaviour regarding health risks and generational cohorts.

Travellers' Risk-Taking Approach and Generational Cohorts

Generational cohorts are individuals born and raised during a specific period and share similar experiences, values, and attitudes (Schewe & Noble, 2000). Each cohort lasts around 20 years until the next one arises (Pendergast, 2009), with similar cultural views, norms, and shared memories (Lebrun et al., 2022). It is one of the most popular consumer segmentation methods. In tourism research, it is important to understand generational cohorts because it allows researchers to identify the unique travel behaviours and preferences of different age groups. Such an understanding can help tourism businesses and destinations to target their marketing efforts better and tailor their offerings to specific generations (Chen & Chou, 2019). To date, travellers' risk perception and risk-taking behaviour associated with health concerns have been a topic of interest among researchers and practitioners in tourism (e.g. Pantano et al., 2022; Villacé-Molinero et al., 2021; Joo et al., 2021). Nevertheless, understanding travellers' risk-taking behaviour across generational cohorts in the context of COVID-19 is insufficient. Research suggests that generational theory has significant practical and theoretical implications for several behavioural phenomena (Šerić et al., 2022). This section focuses on four main generational cohorts to generate insights into travellers' risk-taking behaviour (see Table 8.1): Baby Boomers, Gen X, Gen Y, and Gen Z.

Baby Boomers and Their Risk-Taking Behaviour

Baby Boomers have been defined as individuals born between 1946 and 1964 (Babin et al., 2021) who are between 59 and 77 years old. This generation was steeped in the post-World War II era and experienced general prosperity, including social, political, and economic aspects (Francis & Hoefel, 2018). This experience of prosperity made the

Table 8.1 The difference between generational cohorts (the authors)

	Baby Boomer 1946–1964	Generation X 1965–1980	Generation Y 1981–1995	Generation Z 1996–2010
Context	• Post-World War II era	• Economic uncertainty • Societal uncertainty	• Globalisation • Empiricism	• Digitally native
Characteristics	• Idealistic • Narcissistic • Self-confident • Hedonism • conservative	• Individualistic • Self-reliance • Self-sufficiency • Materialistic	• Optimistic • Team-oriented • High self-expression • Respectful and trusting of authority	• Realistic • Individualistic • Diverse • Tech-savvy • Well-educated
Risk-taking approach	• Risk-avoid	• Risk-averse	• Risk-averse	• Risk-tolerant

members of Baby Boomers idealistic, optimistic, self-confident, and consumption-oriented (Rosen, 2011).

Moreover, according to Rosen (2011), the Baby Boomer generation is characterised by its propensity for taking social and public action against things they deem unjust or unfair.

Several studies have also characterised Baby Boomers as narcissistic, hedonistic, profligate consumers of natural resources, and individualistic (Lissitsa et al., 2022; Lissitsa & Laor, 2021). Regarding the use of technology, the members of Baby Boomers have the lowest level of engagement with information technologies and social media compared with other generations, while their use of the Internet and information technologies has been on the rise (Guerrieri & Bentivegna, 2011). Drawing on the existing literature (Francis & Hoefel, 2018; Lissitsa & Laor, 2021) and above, several common characteristics of Baby Boomers have been summarised as follows: (1) idealistic; (2) narcissistic; (3) self-confident; (4) hedonism; and (5) conservative.

In addition to their consumption behaviour, Baby Boomers were found to have certain characteristics that set them apart from other generations. For instance, Sperazza & Banerjee (2010) have discovered that Boomers emphasise leisure as a way to continue growing rather than just relaxing. They also tend to prioritise quality and value in their purchasing decisions and are strongly inclined towards nostalgic and traditional brands (Ginting et al., 2022). In tourism, this generation has been considered the most valuable market for destinations and travel providers among the other three generations (Li et al., 2013). According to Horneman et al. (2002), affluent seniors travel more often, with greater distances and longer duration than younger generations. Since Baby Boomers have reached retirement age, they have more disposable income and more leisure time to travel (Tiago et al., 2016).

However, in recent years, Baby Boomers have suffered the most negative effects of coronavirus compared with the other generational cohorts due to the pandemic (Wisker, 2022). Previous clinical research has indicated that older adults may be more susceptible to COVID-19 because the infection can exacerbate pre-existing diseases (Armitage & Nellums, 2020). Over 80% of coronavirus deaths in the United States were 65 and older in 2020 (Osama et al., 2020). This fact may enhance Baby Boomers' self-consciousness and death-related anxiety, causing a rise in self-protective actions (Remuzzi & Remuzzi, 2020). As a result, Baby Boomers' travel activity has significantly changed during the pandemic. They stopped travelling for leisure, and the main reasons for their journey were to go to supermarkets, pharmacies, and newspaper stands (Dadashzadeh et al., 2022). Although all generational cohorts experienced this shift, scholars found that older generations react differently, for instance, a decline in Baby Boomers' travel activities that occurred earlier and greater than younger generations (Kabiri et al., 2020). Besides that, a recent study by Lebrun et al. (2022) reveals that Baby Boomers' travel intentions are very different from the other cohorts as they are the most aware of the risks associated with COVID-19.

Generation X and Their Risk-Taking Behaviour

Generation X (Gen X), also known as the baby-bust generation, has been defined as individuals born between 1965 and 1980 (Reisenwitz & Fowler, 2019; Stylos et al., 2021b). This generation grew up with both economic uncertainty (e.g. the stock market crash of 1987 and the recession in 1990) and societal uncertainty (e.g. "latchkey kids") (Goldring & Azab, 2021). Thus, members of Gen X care more about a sense of belonging, security, and feedback (Jurkiewicz, 2000). They are more price-conscious and have a short- term orientation because of their exposure to the crisis (Harari et al., 2022). A previous study has also indicated that Gen X are very independent and rebellious (Peluchette et al., 2013) because most Gen X members grew up in an era in which both of their parents were in the workforce (Lissitsa & Kol, 2016). Given the above, several common characteristics of Gen X have been identified: (1) individualism; (2) self-reliance; (3) self-sufficiency; and (4) materialism.

In a tourism consumption context, scholars have indicated that Gen X has sophisticated consumption behaviour (Lissitsa & Kol, 2016). In particular, Reisenwitz and Iyer (2009) have characterised Gen Xers as more cynical, pessimistic, and less idealistic than previous generations. Consequently, this generation tends to be sceptical of their current situation. Thus, Gen X expect a detailed explanation of services/products' features and why these features are needed before making a purchase decision (Himmel, 2008). They also care about others' options for the products/services. Hence, Gen X are more likely to read reviews and visit more opinion sites than any other generation (Lissitsa & Kol, 2016). Compared to other generations, Generation X has a more family-oriented travel behaviour focussing on family needs. Their decision-making process can be more influenced by reviews and informative content from destinations or travel brands Reisenwitz & Fowler, 2019).

Regarding the risks, previous research has reported that Gen X is more risk-averse than other generations. Reisenwitz and Iyer (2009) reveal that Gen X is a reactive generation with a low-risk tolerance due to their pessimistic nature. During the pandemic, tourism scholars have demonstrated that Gen X has higher conservation values than younger generations (Harari et al., 2022).

Generation Y and Their Risk-Taking Behaviour

Generation Y (Gen Y), or Millennials, was defined as the people born between 1981 and 1995 (Li et al., 2013). According to Benckendorff &

Moscardo (2009), Gen Y grew up in the era of globalisation and empiricism. As such, the members of Gen Y have strong self-centredness and self-actualisation desires and a relatively high level of self-expression (Park & Gursoy, 2012; Luger et al., 2022). Moreover, Pendergast (2009) considered Gen Y the hero generation and characterised them as conventional with respect for authority. Compared with other generations, Gen Y has been 'sheltered' and 'watched over' more and spent more time with structured activities (Benckendorff & Moscardo, 2009). They tend to be influenced by the viewpoints of their family members and social circle more than other generations (Prasad et al., 2019). By reviewing previous generational cohort literature (Francis & Hoefel, 2018; Benckendorff et al., 2009; Stylos et al., 2021a), this study has summarised several common characteristics of Gen Y, as follows: (1) optimistic; (2) team-oriented; (3) high self-expression; and (4) respectful and trusting of authority.

Regarding tourism consumption behaviour, prior studies have indicated that Gen Y has high spending power but is less loyal (Ladhari et al., 2019; Parment, 2013). This generation is more likely to try new services/products and is highly consumption-oriented (Eastman & Liu, 2012). In other words, Gen Y prefers experiential consumption experiences as they are "experience seekers" (Leask et al., 2014). Moreover, Jennings et al. (2009) discovered that members of Gen Y prefer services/products which obtain value from the physical engagement. In comparison to other generational cohorts, Gen Y has unique consumption characteristics. Francis and Hoefel (2018) identified three consumption features as follows: (1) desire for a more integrated experience; (2) relying more on the opinions of their peers and word-of-mouth; and (3) remaining apprehensive and untrusting of commercial activities.

During the pandemic, travellers' behaviours regarding health risks have gained increasing attention from tourism scholars and practitioners. Moskowitz (2019) indicates that Gen Y is the "most risk-averse generation since The Great Depression." More recently, an empirical study by Reisebwitz (2021) reveals that Gen Y tends to avoid risks due to their continual state of safety and security. Their findings also reveal no significant differences in the risk aversion of Gen Y and Gen X during COVID-19.

Generation Z and Their Risk-Taking Behaviour

Generation Z (Gen Z), also known as Centennials (Llopis-Amorós et al., 2019), is defined as people born between 1996 and 2010 (Bassiouni & Hackley, 2014). According to Liu et al. (2022), Gen Z is the first generation to grow up with digital technologies and heavily relies on its use as a digitally native generation. In other words, this generation is tech-savvy, highly educated, and innovative (Priporas et al., 2017). It is projected to be more diverse and connected than other generations (Shatto & Erwin, 2016). As such, Gen Z is more likely to engage in social interaction and co-creating experiences through social media (Skinner et al., 2018). Moreover, Stylos et al. (2021) indicated that Gen Z, the newest generation, tends to become more pragmatic and individualistic. Given the above and prior literature (Francis & Hoefel, 2018), the characteristics of Gen Z can be summarised as follows: (1) realistic; (2) individualistic; (3) diverse; (4) tech-savvy; and (5) well-educated. In recent years, Gen Z is set to become the youngest and largest consumer group (Euromonitor, 2018), accounting for one-third of the world's population (Ameen et al., 2023).

As Gen Z is gradually stepping into the market, tourism and marketing scholars are paying ever more attention to the consumption behaviour of Gen Z. For instance, Abdullah et al. (2018) indicate that Gen Z has more expectations, needs, and new ideas for services/products compared with other generations. Li and Huang (2023) further highlighted that Gen Z is more personalised and focused on the emotional experience. A recent study conducted by Liu et al. (2022) has summarised three consumption behaviour features of Gen Z as follows: (1) prefer customised services/products, (2) prefer products/services that show their individuality and uniqueness, and (3) are more sensitive to social impact. In tourism, travellers' consumption behaviour has significantly changed due to the COVID-19 pandemic in recent years. It is, thus, imperative to understand Gen Z's travel behaviour regarding health risks in the era of post-COVID-19.

Regarding the risk-taking approach, young generations share different travel motivations and personality traits with other generations, such as sensation-seeking and a stronger inclination to engage in physical risk (Stylos et al., 2021). During the pandemic, health risks have become a major concern for travellers. However, Dragin et al. (2022) found that Gen Z travellers' response to health risks (e.g. COVID-19) was consistent concern about non-health risks. Their results reveal that Gen Z denies health risks and expects some travel benefits (e.g. accumulating some savings) from the crisis. Compared with the previous generation, Gen Z is more risk-tolerant (Williams & Baláž, 2014). In other words, young generations are more likely to take risks during travel.

Influence of Risk Perception on Tourism

In recent years, COVID-19 has posed a massive obstacle to visitors' on-site travel and has significantly impacted destination growth (Sigala, 2020; Zhang et al., 2023). This section discusses the impacts of risk perception on tourists' decision-making and destination image among different generational cohorts.

Risk Perception on Destination Image

The concept of "destination image" has been defined as an individual's mental images of a place, including cognitive, affective, and conative attributes (Beerli & Martin, 2004). The cognitive dimension refers to what a person thinks or knows about the features or qualities of a tourist destination (San Martín & Del Bosque, 2008), while the affective part involves the individual's feelings about the place (Hallmann et al., 2015). The conative components refer to how an individual responds to this information (Pike & Ryan, 2004). Recent research suggests that the trend in destination images was towards user-generated content due to the explosive growth of social media, and sources of destination information became more diverse and accessible (Nguyen & Tong, 2023). In tourism, the destination image is the traveller's overall impression of a certain destination (Stylidis, 2022) or the sum of tourists' beliefs, knowledge, impressions, and affective image of a specific destination (Yang et al., 2022). It has long been considered essential in predicting travel behaviour intention and improving the attractiveness and market competitiveness of destinations (Rasoolimanesh et al., 2021).

Tourism, as a type of service industry, is intangible and experiential (Smith, 1994) and is highly susceptible to natural disasters (e.g. COVID-19) (Cambra-Fierro et al., 2022; Pantano et al., 2020; Fong et al., 2021; Karabulut et al., 2020) and other disasters (e.g. terrorist attacks; earthquakes) (Cruz-Milan et al., 2016). Such susceptibility might damage a destination's image. In particular, prior literature has found that natural disasters raise visitors' sense of psychological risk and, as a result, damage the destination's image (Li et al., 2018). Moreover, Polas et al. (2022) highlighted that travellers usually evaluate the attractiveness of a destination based on their perceptions instead of reality. Thus, the impact of travellers' perceived risk on destination image has become an increasing concern among tourism scholars in recent years (Liu et al., 2022; Yang et al., 2022).

Prior research suggests that destination images are significantly influenced by travellers' perceptions of safety and risk (Rasoolimanesh et al., 2021). Indeed, risk perceptions have been considered a part of destination image as it is a pertinent factor in the destinations' overall image (Promsivapallop & Kannaovakun, 2017). During the pandemic, people are not permitted to travel freely. They may be more exposed to media news than usual so that the image of the destination is conveyed by both mass and social media, especially in the early era of COVID-19 (Rasoolimanesh et al., 2021). Travellers' risk perception can also be shaped by looking at others' online reactions and fears towards pandemics (Zheng et al., 2021). In this vein, several studies have shown that media coverage plays a significant role in tourists' risk perception of a geographic area (Sonmez & Graefe, 1998), affecting the destination's image (Schroeder et al., 2013) and travellers' decision-making. As mentioned, individuals perceive risk differently based on their generational cohorts. Not surprisingly, the impact of travellers' risk perceptions on destination image differs among different generations.

Risk Perception on Decision-Making

The term "decision-making" has been used to describe the processes that lead to action with the desired result in mind (Yates & De Oliveira, 2016). People collect information and analyse the advantages and disadvantages of each possible result, eventually choosing the best suitable one from a range of alternatives to accomplish their goal (Smallman & Moore, 2010). Previous literature illustrates that travellers' decision-making builds on how they expect to perceive risks, whether general or specific to their trip (Pappas & Glyptou, 2021). Travellers' risk perception refers to the uncertainty and the likelihood of negative outcomes from purchasing tourism products/services (Matiza, 2020). Tourism scholars have widely acknowledged it as one of the most important aspects of decision-making in tourism (e.g. Wang et al., 2022; Müller & Wittmer, 2023).

Moreover, travellers' decision-making process involves cognitive contextual facts and emotional opinions, with the latter vulnerable to many intangible factors (Williams & Baláž, 2014). Risk perceptions, thus, are more of a contextualisation of the probability and size of the negative repercussions of an event rather than the factual evaluation of the risk (Pappas & Glyptou, 2021). Recent studies on COVID-19 indicate how dramatically health-related risks may change travellers' travel intentions and decision-making. Within the decision-making process, people evaluate the boundaries of risk and uncertainty and maximise utility while minimising loss based on their beliefs, values, and experiences (Karl, 2018). The pandemic made travellers more ethnocentric, xenophobic, oversensitive to crowds and likely to opt for group trips and insurance (Karl et al., 2021). In the era of COVID-19, when presented with risks, travellers may either postpone their travel plans and reconsider their place of choice or cancel their visits entirely (Matiza, 2020; Zenker & Kock, 2020).

In tourism, health risks (e.g. COVID-19) related studies have indicated that age or generation significantly moderates travellers' behaviour intentions and decision-making in the context of the pandemic (Lebrun et al., 2022). This moderation is because the risk of serious sickness with COVID-19 rises with age, with the highest risk among older people. An increasing number of empirical studies have explored the travellers' decision-making associated with health risks from a generational cohort perspective. For instance, Abraham et al. (2020) examined the relationship between the perceived risk associated with COVID-19 pandemic and their future travel intentions to China in different generational cohorts. Their findings reveal significant differences across generational cohorts. Similarly, Lebrun et al. (2022) studied the generational difference in the influence of health risks on French tourists' travel intentions during the pandemic. Their results also demonstrate significant differences in travel intentions across generations in France, with 66% of Generation X, 55% of Generation Y, 50% of Baby Boomers, and 44% of Generation Z.

Conclusions and Recommendations

In conclusion, this chapter provides valuable insights into four different generational cohorts' risk perceptions and travel behaviour in the context of the pandemic. To better understand the relationship among risk perception, travel behaviour, destination image, and decision-making across generations, we develop a framework as shown in Fig. 8.1. Specifically, risk perception has been identified as an important determinant of travel behaviour. Tourists usually avoid travel destinations and activities they perceive as risky (Wang et al., 2019). Moreover, people in different generations have different perceptions regarding health risks because each generation cohort has its characteristics, consumer behaviour, and social, and economic environment in which they grow up. Along these lines, this study has found that the influences of travellers' risk perception on destination image and decision-making differ across generations.

Future Agenda

Although much has been written about risk perceptions and generational cohorts in previous studies, the tourism literature on different generational cohorts' risk-taking behaviours associated with health-related risks

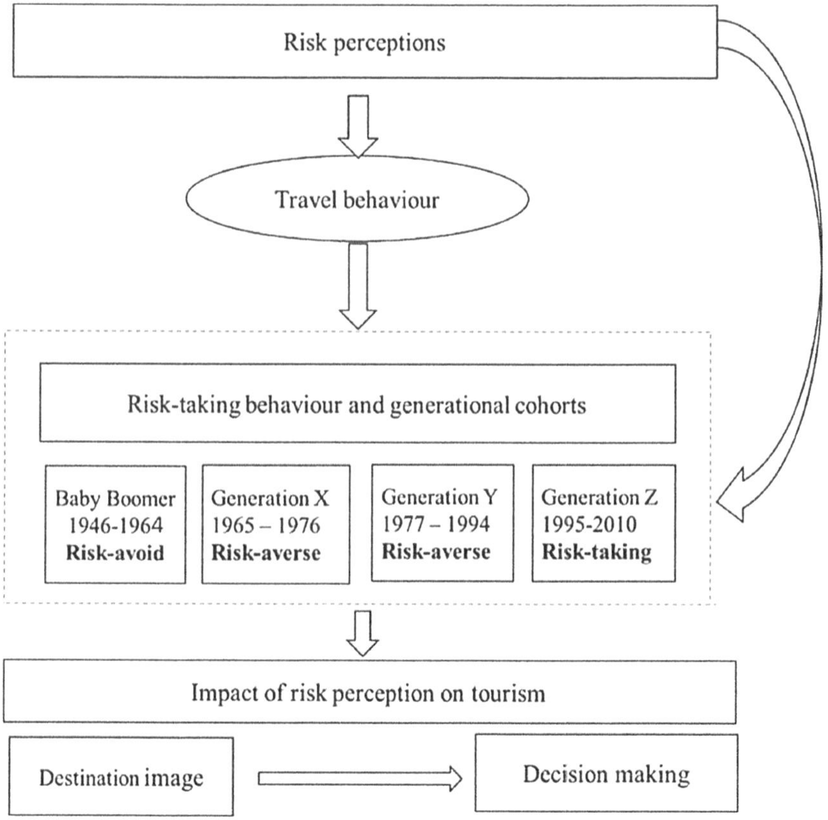

Fig. 8.1 Impact of risk perception on travellers' behaviour across different generations. (Source: Authors)

in the context of the COVID-19 pandemic remains limited. A discussion of the future agenda for health-related risks in tourism is thus valuable for both tourism scholars and practitioners. This study sets up a future agenda which identifies the promising directions for future tourism research in risk perceptions based on four different generation cohorts, as summarised in Table 8.2, including Baby Boomers, Generation X, Generation Y, and Generation Z. This chapter also suggests some recommendations for tourism practitioners to target their consumers and improve their competitiveness in the era of post-COVID-19. Details will be discussed in the next subsections.

Table 8.2 Future agenda for tourism scholars and practitioners

Generational cohorts	Academics	Practitioners
Baby Boomers	• To understand how to meet Baby Boomers' needs in a post-COVID-19 era • To empirically investigate the factors that affect Baby Boomers' perceived destination image after COVID-19	• To adjust marketing strategies to attract Baby Boomers • To include Baby oomers' risk perception in the destination image formation process
Generation X	• To investigate Gen X's travel behaviour changes during the pandemic • To explore Gen X's risk-taking behaviour in different countries	• To emphasise the competitive price of products/services • To encourage customer reviews
Generation Y	• To test the difference in risk-taking behaviour between Gen X and Gen Y in the post-COVID-19 era • To empirically investigate how Generation Y's travel preferences and behaviours have changed due to the pandemic	• Be transparent and communicate any changes in policy or procedures that may affect their travel experience • Promote safety and hygiene measures
Generation Z	• To understand the long-term effects of the pandemic on Generation Z's travel behaviour	• Create unique and authentic experiences • Utilise digital platforms

Source: Authors

Baby Boomers

Senior travellers have more spending power and leisure time to travel compared with other generations (Tiago et al., 2016). Thus, several studies have indicated that Baby Boomers will likely play an important role in the tourism industry's revival after the COVID-19 epidemic (Patterson & Balderas-Cejudo, 2022; Sharma et al., 2022). Accordingly, future tourism studies on destination management and tourism development should further investigate the Baby boomer's risk perception and travel behaviour in the era of post-COVID-19. Moreover, Patterson and

Balderas-Cejudo (2022) call for more empirical studies on travellers' health concerns focussing on generational cohorts, such as Baby Boomers. Hence, future scholars could empirically investigate the factors that affect Baby Boomers' perceived destination image in the post-COVID-19 era. Such an investigation would provide deeper insights into tourism recovery after the pandemic.

For tourism practitioners, we suggest that tourism and destination managers should adapt their marketing strategies to target Baby Boomers travellers. In particular, their marketing and communication efforts should target Baby Boomers and their specific interests, health concerns, and needs. Moreover, travellers' destination image perceptions significantly influence their decision-making and travel intention. Practitioners should consider Baby Boomers' risk perceptions in the destination image formation process, as health and safety are their main concerns.

Generation X

By drawing on the existing generational research in tourism, Gen X shares similarity and difference with other generations. To help tourism marketers better target Gen X, future research needs to ascertain the impact of COVID-19 on the travel preferences and behaviours of Generation X, including changes in destination choices, travel modes, and budget considerations. Moreover, further testing Gen X's risk-taking behaviour in different countries would be interesting.

As mentioned, members of Gen X are more price-conscious and have a short-term orientation because of their exposure to the crisis (Harari et al., 2022). Thus, tourism and destination managers should emphasise the competitive price of products/services when targeting Gen X travellers. Moreover, Gen X is more sceptical and cares about others' options for products/services (Lissitsa & Kol, 2016). Destination service providers should put more effort into improving online word of mouth, for instance, to encourage travellers to write a review about destination risk management through some promotion tools.

Generation Y

According to Reisebwitz (2021), there are no significant differences in the risk aversion of Gen Y and Gen X during COVID-19. However, travellers' risk perception and travel behaviour towards COVID-19 have changed after the pandemic. Thus, it would be interesting for future studies to test the difference in risk-taking behaviour between Gen X and Gen Y in the post-COVID-19 era. Moreover, empirically investigating how Generation Y's travel preferences and behaviours have changed due to the pandemic can also guide tourism managers to adapt their marketing strategies for attracting Gen Y customers.

Since Gen Y remains apprehensive and untrusting of commercial activities, tourism managers need to be transparent and communicate any changes in policy or procedures that may affect their travel experience. Moreover, previous studies have found that Gen Y has been the most risk-averse generation since The Great Depression (Moskowitz, 2019). Thus, tourism and destination managers should promote safety and hygiene measures in the post-COVID-19 era, such as increased cleaning and social distancing protocols, to assure them of the safety of their destination.

Generation Z

According to Robinson and Schänzel (2019), understanding Gen Z travellers will be key to predicting how tourism trends could evolve in the next few years and how these could influence worldwide tourism. Thus, future researchers must understand the long-term effects of the pandemic on the travel and tourism industry, specifically in terms of Generation Z travellers. Moreover, future research should adopt experimental design and other quantitative methods to explicitly examine Gen Z's risk-taking behaviour after the pandemic by comparing three other generational cohorts.

Generation Z values experience over material possessions, so tourism managers should focus on creating unique and authentic experiences that align with their interests and values. Moreover, Generation Z is tech-savvy and more likely to engage in social interaction and co-create experiences through social media. Tourism managers should ensure their website is user-friendly and mobile-optimised and have active social media accounts to engage with potential customers.

Recommended Further Reading

Li, J., Nguyen, T. H. H., & Coca-Stefaniak, J. A. (2020). Coronavirus impacts on post- pandemic planned travel behaviors. *Annals of Tourism Research.*

Liu, J., Wang, C., Zhang, T., & Qiao, H. (2022). Delineating the effects of social media marketing activities on generation Z travel behaviors. *Journal of Travel Research*, 00472875221106394.

Neuburger, L., & Egger, R. (2021). Travel risk perception and travel behavior during the COVID-19 pandemic 2020: A case study of the DACH region. *Current Issues in Tourism, 24*(7), 1003–1016.

Pantano, E., Scarpi, D., Vannucci, V., Bilotta, E., & Pantano, P. (2022). Probability-density risk maps for tourism during emergencies. *Annals of Tourism Research, 92*(C).

Pantano, E., Vannucci, V., & Marikyan, D. (in press). Gratification in change of privacy? The response of four consumers' generational cohort toward facial recognition technology in retail settings. *Journal of Consumer Behavior.*

Stylos, N. (2022). An integrated duality theory framework (IDTF): Marking pathways for consumer decision-making researchers in the hospitality and tourism industry. *International Journal of Contemporary Hospitality Management, 34*(7), 2597–2619.

Stylos, N., Rahimi, R., Okumus, B., & Williams, S. (2021). *Generation Z marketing and management in tourism and hospitality*. Springer.

Stylos, N., & Vassiliadis, C. A. (2023). Gamification design: Toward developing image perception scales for generation Z consumers. *International Journal of Contemporary Hospitality Management.* https://doi.org/10.1108/IJCHM-12-2022-1616

Zheng, D., Luo, Q., & Ritchie, B. W. (2021). Afraid to travel after COVID-19? Self-protection, coping and resilience against pandemic' travel fear'. *Tourism Management, 83*, 104261.

References

Abdullah, A., Ismail, M. M., & Albani, A. (2018). At-risk generation Z: values, talents and challenges. *International Journal of Asian Social Science, 8*(7), 373–378.

Abraham, V., Bremser, K., Carreno, M., Crowley-Cyr, L., & Moreno, M. (2020). Exploring the consequences of COVID-19 on tourist behaviors: Perceived travel risk, animosity and intentions to travel. *Tourism Review.*

Ameen, N., Hosany, S., & Taheri, B. (2023). Generation Z's psychology and new-age technologies: Implications for future research. *Psychology & Marketing, 40*(10), 2029–2040.

Armitage, R., & Nellums, L. B. (2020). COVID-19 and the consequences of isolating the elderly. *The Lancet Public Health, 5*(5), e256.

Babin, B. J., Zhuang, W., & Borges, A. (2021). Managing service recovery experience: effects of the forgiveness for older consumers. *Journal of Retailing and Consumer Services, 58*, 102222.

Bae, S. Y., & Chang, P. J. (2021). The effect of coronavirus disease-19 (COVID-19) risk perception on behavioral intention towards 'untact' tourism in South Korea during the first wave of the pandemic (March 2020). *Current Issues in Tourism, 24*(7), 1017–1035.

Bassiouni, D. H., & Hackley, C. (2014). Generation Z' children's adaptation to digital consumer culture: A critical literature review. *Journal of Customer Behavior, 13*(2), 113–133.

Beerli, A., & Martin, J. D. (2004). Factors influencing destination image. *Annals of Tourism Research, 31*(3), 657–681.

Benckendorff, P., & Moscardo, G. (2009). Understanding Generation-Y tourists: managing the risk and change associated with a new emerging market. In *Tourism and generation Y* (pp. 38–46). CABI.

Brewer, N. T., & Fazekas, K. I. (2007). Predictors of HPV vaccine acceptability: A theory- informed, systematic review. *Preventive Medicine, 45*(2–3), 107–114.

Cambra-Fierro, J., Fuentes-Blasco, M., Gao, L. X., Melero-Polo, I., & Trifu, A. (2022). The influence of communication in destination imagery during COVID-19. *Journal of Retailing and Consumer Services, 64*, 102817.

Chen, C. F., & Chou, S. H. (2019). Antecedents and consequences of perceived coolness for generation Y in the context of creative tourism-a case study of the pier 2 art Center in Taiwan. *Tourism Management, 72*, 121–129.

Chua, B. L., Al-Ansi, A., Lee, M. J., & Han, H. (2021). Impact of health risk perception on avoidance of international travel in the wake of a pandemic. *Current Issues in Tourism, 24*(7), 985–1002.

Cruz-Milán, O., Simpson, J. J., Simpson, P. M., & Choi, W. (2016). Reassurance or reason for concern: Security forces as a crisis management strategy. *Tourism Management, 56*, 114–125.

Dadashzadeh, N., Larimian, T., Levifve, U., & Marsetič, R. (2022). Travel behavior of vulnerable social groups: Pre, during, and post COVID-19 pandemic. *International Journal of Environmental Research and Public Health, 19*(16), 10065.

Dolnicar, S., & Zare, S. (2020). COVID-19 and Airbnb–disrupting the disruptor. *Annals of Tourism Research, 83*, 102961.

Dragin, A. S., Majstorović, N., Janičić, B., Mijatov, M. B., & Stojanović, V. (2022). Clusters of generation Z and travel risks perception: Constraining vs push–pull factors. In *The emerald handbook of destination recovery in tourism and hospitality* (pp. 375–395). Emerald Publishing Limited.

Duro, J. A., Perez-Laborda, A., Turrion-Prats, J., & Fernández-Fernández, M. (2021). COVID-19 and tourism vulnerability. *Tourism Management Perspectives, 38*, 100819.

Eastman, J. K., & Liu, J. (2012). The impact of generational cohorts on status consumption: An exploratory look at generational cohort and demographics on status consumption. *Journal of consumer marketing.*

Euromonitor. (2018). *Generation Z: The next wave of consumers.* September 20. http://www.portal.euromonitor.com.ezproxy.mdx.ac.uk/portal/analysis/tab/

Floyd, D. L., Prentice-Dunn, S., & Rogers, R. W. (2000). A meta-analysis of research on protection motivation theory. *Journal of Applied Social Psychology, 30*(2), 407–429.

Foo, L. P., Chin, M. Y., Tan, K. L., & Phuah, K. T. (2021). The impact of COVID-19 on the tourism industry in Malaysia. *Current Issues in Tourism, 24*(19), 2735–2739.

Fong, L. H. N., Law, R., & Ye, B. H. (2021). Outlook of tourism recovery amid an epidemic: Importance of outbreak control by the government. *Annals of Tourism Research, 86*, 102951.

Francis, T., & Hoefel, F. (2018). True Gen': Generation Z and its implications for companies. *McKinsey & Company, 12.*

Ginting, Y. M., Sinaga, A. M. R., & Nainggolan, R. D. (2022). Analysis the influence of digital marketing, product differentiation, customer value, service quality to purchase decision and repurchase intention of millennial generation at Shopee online shop. *SEIKO: Journal of Management & Business, 4*(3), 134–153.

Goldring, D., & Azab, C. (2021). New rules of social media shopping: Personality differences of US gen Z versus gen X market mavens. *Journal of Consumer Behavior, 20*(4), 884–897.

Guerrieri, P., & Bentivegna, S. (Eds.). (2011). *The economic impact of digital technologies: Measuring inclusion and diffusion in Europe.* Edward Elgar Publishing.

Haddock, C. (1993). *Managing risks in outdoor activities.* New Zealand Mountain Safety Council.

Hallmann, K., Zehrer, A., & Müller, S. (2015). Perceived destination image: An image model for a winter sports destination and its effect on intention to revisit. *Journal of Travel Research, 54*(1), 94–106.

Harari, T. T. E., Sela, Y., & Bareket-Bojmel, L. (2022). Gen Z during the COVID-19 crisis: A comparative analysis of the differences between gen Z and gen X in resilience, values and attitudes. *Current Psychology*, 1–10.

Himmel, B. (2008). Different strokes for different generations. *Rental Product News, 30*(7), 42–46.

Hoque, A., Shikha, F. A., Hasanat, M. W., Arif, I., & Hamid, A. B. A. (2020). The effect of coronavirus (COVID-19) in the tourism industry in China. *Asian Journal of Multidisciplinary Studies, 3*(1), 52–58.

Horneman, L., Carter, R. W., Wei, S., & Ruys, H. (2002). Profiling the senior traveller: An Australian perspective. *Journal of Travel Research, 41*(1), 23–37.

Im, J., Kim, J., & Choeh, J. Y. (2021). COVID-19, social distancing, and risk-averse actions of hospitality and tourism consumers: A case of South Korea. *Journal of Destination Marketing & Management, 20*, 100566.

Jennings, G., Cater, C., Lee YoungSook, L. Y., Ollenburg, C., Ayling, A., & Lunny, B. (2009). Generation Y: Perspectives of quality in youth adventure travel experiences in an Australian backpacker context. In *Tourism and generation Y* (pp. 58–72). Cabi.

Joo, D., Xu, W., Lee, J., Lee, C. K., & Woosnam, K. M. (2021). Residents' perceived risk, emotional solidarity, and support for tourism amidst the COVID-19 pandemic. *Journal of Destination Marketing & Management, 19*, 100553.

Jurkiewicz, C. L. (2000). Generation X and the public employee. *Public personnel management, 29*(1), 55–74.

Kabiri, A., Darzi, A., Zhou, W., Sun, Q., & Zhang, L. (2020). How different age groups responded to the COVID-19 pandemic in terms of mobility behaviors: A case study of the United States. *arXiv preprint arXiv:2007.10436.*

Karl, M. (2018). Risk and uncertainty in travel decision-making: Tourist and destination perspective. *Journal of Travel Research, 57*(1), 129–146.
Karl, M., Kock, F., Ritchie, B. W., & Gauss, J. (2021). Affective forecasting and travel decision-making: An investigation in times of a pandemic. *Annals of Tourism Research, 87*, 103139.
Karabulut, G., Bilgin, M. H., Demir, E., & Doker, A. C. (2020). How pandemics affect tourism: International evidence. *Annals of Tourism Research, 84*, 102991.
Kim, E. E. K., Seo, K., & Choi, Y. (2022). Compensatory travel post COVID-19: Cognitive and emotional effects of risk perception. *Journal of Travel Research, 61(8)*, 1895–1909.
Ladhari, R., Gonthier, J., & Lajante, M. (2019). Generation Y and online fashion shopping: Orientations and profiles. *Journal of Retailing and Consumer Services, 48*, 113–121.
Law, R. (2006). The perceived impact of risks on travel decisions. *International Journal of Tourism Research, 8*(4), 289–300.
Leask, A., Fyall, A., & Barron, P. (2014). Generation Y: An agenda for future visitor attraction research. *International Journal of Tourism Research, 16*(5), 462–471.
Lebrun, A. M., Corbel, R., & Bouchet, P. (2022). Impacts of COVID-19 on travel intention for summer 2020: A trend in proximity tourism mediated by an attitude towards COVID-19. *Service Business, 16*(3), 469–501.
Li, C., & Huang, X. (2023). How does COVID-19 risk perception affect wellness tourist intention: Findings on Chinese generation Z. *Sustainability, 15*(1), 141.
Li, F., Wen, J., & Ying, T. (2018). The influence of crisis on tourists' perceived destination image and revisit intention: An exploratory study of Chinese tourists to North Korea. *Journal of Destination Marketing & Management, 9*, 104–111.
Li, J., Nguyen, T. H. H., & Coca-Stefaniak, J. A. (2020). Coronavirus impacts on post- pandemic planned travel behaviors. *Annals of Tourism Research.*
Li, X., Li, X. R., & Hudson, S. (2013). The application of generational theory to tourism consumer behavior: An American perspective. *Tourism Management, 37*, 147–164.
Lissitsa, S., & Kol, O. (2016). Generation X vs. generation Y–A decade of online shopping. *Journal of Retailing and Consumer Services, 31*, 304–312.

Lissitsa, S., & Laor, T. (2021). Baby boomers, generation X and generation Y: Identifying generational differences in effects of personality traits in on-demand radio use. *Technology in Society, 64*, 101526.

Lissitsa, S., Zychlinski, E., & Kagan, M. (2022). The silent generation vs baby boomers: Socio-demographic and psychological predictors of the "gray" digital inequalities. *Computers in Human Behavior, 128*, 107098.

Liu, J., Wang, C., Zhang, T., & Qiao, H. (2022). Delineating the effects of social media marketing activities on generation Z travel behaviors. *Journal of Travel Research*, 00472875221106394.

Llopis-Amorós, M. P., Gil-Saura, I., Ruiz-Molina, M. E., & Fuentes-Blasco, M. (2019). Social media communications and festival brand equity: Millennials vs Centennials. *Journal of Hospitality and Tourism Management, 40*, 134–144.

Luger, M., Hofer, K. M., & Floh, A. (2022). Support for corporate social responsibility among generation Y consumers in advanced versus emerging markets. *International Business Review, 31*(2), 101903.

Matiza, T. (2020). Post-COVID-19 crisis travel behavior: Towards mitigating the effects of perceived risk. *Journal of Tourism Futures, 8*(1), 99–108.

Moskowitz, D. (2019, June 25). *Are Millennials risk averse or risk takers? Investopedia*. https://www.investopedia.com/articles/investing/070815/ar

Müller, A., & Wittmer, A. (2023). The choice between business travel and video conferencing after COVID-19–Insights from a choice experiment among frequent travelers. *Tourism Management, 96*, 104688.

Neuburger, L., & Egger, R. (2021). Travel risk perception and travel behavior during the COVID-19 pandemic 2020: A case study of the DACH region. *Current Issues in Tourism, 24*(7), 1003–1016.

Nguyen, T. T. T., & Tong, S. (2023). The impact of user-generated content on intention to select a travel destination. *Journal of Marketing Analytics, 11*(3), 443–457.

Osama, T., Pankhania, B., & Majeed, A. (2020). Protecting older people from COVID-19: Should the United Kingdom start at age 60? *Journal of the Royal Society of Medicine, 113*(5), 169–170.

Pantano, E., Pizzi, G., Scarpi, D., & Dennis, C. (2020). Competing during a pandemic? Retailers' ups and downs during the COVID-19 outbreak. *Journal of Business Research, 116*, 209–213.

Pantano, E., Scarpi, D., Vannucci, V., Bilotta, E., & Pantano, P. (2022). Probability-density risk-maps for tourism during emergencies. *Annals of Tourism Research, 92*(C).

Pappas, N., & Glyptou, K. (2021). Accommodation decision-making during the COVID-19 pandemic: Complexity insights from Greece. *International Journal of Hospitality Management, 93*, 102767.
Patterson, I., & Balderas-Cejudo, A. (2022). Baby boomers and their growing interest in spa and wellness tourism. *International Journal of Spa and Wellness, 5*(3), 237–249.
Park, J., & Gursoy, D. (2012). Generation effects on work engagement among US hotel employees. *International Journal of Hospitality Management, 31*(4), 1195–1202.
Parment, A. (2013). Generation Y vs. baby boomers: Shopping behavior, buyer involvement and implications for retailing. *Journal of Retailing and Consumer Services, 20*(2), 189–199.
Peluchette, J. V. E., Kovanic, N., & Partridge, D. (2013). Helicopter parents hovering in the workplace: What should HR managers do? *Business Horizons, 56*(5), 601–609.
Pendergast, D. (2009). Getting to know the Y generation. In *Tourism and generation Y* (pp. 1–15). Wallingford.
Pike, S., & Ryan, C. (2004). Destination positioning analysis through a comparison of cognitive, affective, and conative perceptions. *Journal of travel research, 42(4)*, 333–342.
Polas, M. R. H., Saha, R. K., & Tabash, M. I. (2022). How does tourist perception lead to tourist hesitation? Empirical evidence from Bangladesh. *Environment, Development and Sustainability, 24*(3), 3659–3686.
Prasad, S., Garg, A., & Prasad, S. (2019). Purchase decision of generation Y in an online environment. *Marketing Intelligence & Planning.*
Priest, S. (1990). The adventure experience paradigm. In A. Miles & S. Priest (Eds.), *Adventure recreation* (pp. 157–162). Venture Publishing.
Priporas, C. V., Stylos, N., & Fotiadis, A. K. (2017). Generation Z consumers' expectations of interactions in smart retailing: A future agenda. *Computers in Human Behavior, 77*, 374–381.
Promsivapallop, P., & Kannaovakun, P. (2017). A comparative assessment of destination image, travel risk perceptions and travel intention by young travellers across three ASEAN countries: A study of German students. *Asia Pacific Journal of Tourism Research, 22*(6), 634–650.
Rasoolimanesh, S. M., Seyfi, S., Rastegar, R., & Hall, C. M. (2021). Destination image during the COVID-19 pandemic and future travel behavior: The moderating role of past experience. *Journal of Destination Marketing & Management, 21*, 100620.

Reisebwitz, T. H. (2021). Difference in generation Y and generation Z: Implications for marketers. *Marketing Management Journal, 31*(2).

Reisenwitz, T. H., & Iyer, R. (2009). Differences in generation X and generation Y: Implications for the organization and marketers. *Marketing Management Journal, 19*(2).

Reisenwitz, T. H., & Fowler, J. G. (2019). Information sources and the tourism decision- making process: An examination of generation X and generation Y consumers. *Global Business Review, 20*(6), 1372–1392.

Reisinger, Y., & Mavondo, F. (2005). Travel anxiety and intentions to travel internationally: Implications of travel risk perception. *Journal of Travel Research, 43*(3), 212–225.

Remuzzi, A., & Remuzzi, G. (2020). COVID-19 and Italy: What next? *The Lancet, 395*(10231), 1225–1228.

Ren, M., Park, S., Xu, Y., Huang, X., Zou, L., Wong, M. S., & Koh, S. Y. (2022). Impact of the COVID-19 pandemic on travel behavior: A case study of domestic inbound travelers in Jeju, Korea. *Tourism Management, 92*, 104533.

Robinson, V. M., & Schänzel, H. A. (2019). A tourism inflex: Generation Z travel experiences. *Journal of Tourism Futures.*

Rosen, L. (2011). *Teaching the iGeneration.*

San Martín, H., & Del Bosque, I. A. R. (2008). Exploring the cognitive–affective nature of destination image and the role of psychological factors in its formation. *Tourism Management, 29*(2), 263–277.

Schewe, C. D., & Noble, S. M. (2000). Market segmentation by cohorts: the value and validity of cohorts in America and abroad. *Journal of Marketing Management, 16*(1–3), 129–142.

Schroeder, A., Pennington-Gray, L., Kaplanidou, K., & Zhan, F. (2013). Destination risk perceptions among US residents for London as the host city of the 2012 Summer Olympic Games. *Tourism Management, 38*, 107–119.

Šerić, M., Došen, Đ. O., & Mikulić, J. (2022). Antecedents and moderators of positive word- of-mouth communication among tourist destination residents during the COVID-19 pandemic. *Current Issues in Tourism*, 1–18.

Sharma, R., Singh, G., & Pratt, S. (2022). Exploring travel envy and social return in domestic travel: A cross-generational analysis. *Journal of Travel & Tourism Marketing, 39*(1), 58–72.

Shatto, B., & Erwin, K. (2016). Moving on from millennials: Preparing for generation Z. *The Journal of Continuing Education in Nursing, 47*(6), 253–254.

Sigala, M. (2020). Tourism and COVID-19: Impacts and implications for advancing and resetting industry and research. *Journal of Business Research, 117*, 312–321.

Skinner, H., Sarpong, D., & White, G. R. (2018). Meeting the needs of the Millennials and Generation Z: gamification in tourism through geocaching. *Journal of Tourism Futures, 4(1)*, 93–104.

Smallman, C., & Moore, K. (2010). PROCESS STUDIES OF TOURISTS'DECISION-MAKING. *Annals of Tourism Research, 37*(2), 397–422.

Smith, S. L. (1994). The tourism product. *Annals of Tourism Research, 21*(3), 582–595.

Sönmez, S. F., & Graefe, A. R. (1998). Influence of terrorism risk on foreign tourism decisions. *Annals of tourism research, 25*(1), 112–144.

Stylos, N., Fotiadis, A. K., Shin, D. D., & Huan, T. C. T. (2021a). Beyond smart systems adoption: Enabling diffusion and assimilation of smartness in hospitality. *International Journal of Hospitality Management, 98*. https://doi.org/10.1016/j.ijhm.2021.103042

Stylos, N., Rahimi, R., Okumus, B., & Williams, S. (2021b). *Generation Z marketing and management in tourism and hospitality*. Springer.

Stylidis, D. (2022). Exploring resident–tourist interaction and its impact on tourists' destination image. *Journal of Travel Research, 61*(1), 186–201.

UNWTO. (2020). *New data shows the impact of COVID-19 on tourism as UNWTO calls for responsible restart of the sector*. Retrieved January 13, 2022, from https://www.unwto.org/news/new-data-shows-impact-of-COVID-19-on-tourism

UNWTO. (2022). *Tourism grows 4% in 2021 but remains far below pre-pandemic levels*. Retrieved January 13, 2022, from https://www.unwto.org/news/tourism-grows-4-in-2021-but-remains-far-below-pre-pandemic-levels

Wang, J., Liu-Lastres, B., Ritchie, B. W., & Mills, D. J. (2019). Travellers' self-protection against health risks: An application of the full protection motivation theory. *Annals of Tourism Research, 78*, 102743.

Wang, L. H., Yeh, S. S., Chen, K. Y., & Huan, T. C. (2022). Tourists' travel intention: Revisiting the TPB model with age and perceived risk as moderator and attitude as mediator. *Tourism Review*.

Williams, A. M., & Baláž, V. (2014). Mobility, risk tolerance and competence to manage risks. *Journal of Risk Research, 17*(8), 1061–1088.

Wisker, Z. L. (2022). Marketing mobile payment to baby boomers during COVID-19 pandemic: The role of emotional appeal, advertising creativity and perceived value. *Journal of Marketing Communications*, 1–23.
Yang, S., Isa, S. M., & Ramayah, T. (2022). How are destination image and travel intention influenced by misleading media coverage? Consequences of COVID-19 outbreak in China. *Vision, 26*(1), 80–89.
Yates, J. F., & De Oliveira, S. (2016). Culture and decision making. *Organizational behavior and human decision processes, 136*, 106–118.
Yu, M., Cheng, M., Yang, L., & Yu, Z. (2022). Hotel guest satisfaction during COVID-19 outbreak: The moderating role of crisis response strategy. *Tourism Management, 93*, 104618.
Zenker, S., & Kock, F. (2020). The coronavirus pandemic–A critical discussion of a tourism research agenda. *Tourism Management, 81*, 104164.
Zheng, D., Luo, Q., & Ritchie, B. W. (2021). Afraid to travel after COVID-19? Self-protection, coping and resilience against pandemic' travel fear'. *Tourism Management, 83*, 104261.
Zhang, S., Sun, T., & Lu, Y. (2023). The COVID-19 pandemic and tourists' risk perceptions: Tourism policies' mediating role in sustainable and resilient recovery in the New Normal. *Sustainability, 15*(2), 1323.

9

GenZers as Marketing Managers

Orhan Can Yılmazdoğan, Cemile Ece, and Efnan Ezenel

Introduction

The idea of generation sheds light on the attributes of people who lived during a particular era by explaining how societies that experienced similar experiences, eras, and systems throughout human history share common behaviors, values, and personality traits (Arslan & Staub, 2015; Bellou et al., 2018). According to Mackay (1997), generational differences arise from how people perceive and respond to recurring events and shifting environmental situations. These differences are also seen in how people conduct themselves in the workplace.

Different ways have been used in business life between generations, particularly with the innovations and technological improvements in the previous 30 years. Many changes have occurred from the X generation, who adopted an authoritarian approach, to the Generation Z, who were born in technology and identified with entrepreneurship. These

O. C. Yılmazdoğan (✉) • C. Ece • E. Ezenel
Eskişehir Osmangazi University, Eskişehir, Turkey
e-mail: cyilmazdogan@ogu.edu.tr

N. Stylos et al. (eds.), *Contemporary Marketing Management for Tourism and Hospitality*, https://doi.org/10.1007/978-3-031-65049-9_9

modifications have altered organizational working practices and managerial procedures (Joshi et al., 2011). Companies should develop new marketing and management approaches for GenZ, which will have new target markets as well as new employees. Businesses must quickly adjust to the new era and new masses, especially in a sector as dynamic as tourism. Because it is anticipated that GenZ, often known as the "Digital Natives," will fundamentally alter marketing knowledge as they enter the workforce in the future (Williams & Page, 2011).

In this book chapter, the role of GenZers as marketing managers, who are or will be working in the tourism sector in the future, has been examined. As a marketing manager, GenZ's expectations from working life, working principles, and self-efficacy features are revealed.

Generations

Although the differences between generations are shaped according to the structure of the period, historical events, and technological development, the fact that there is information transfer and communication between generations should not be ignored (Alwin & McCammon, 2007). While the generation is defined as a human community that shares a certain culture in a certain period, has a collective cultural space, and a collective memory, it also includes an intergenerational transfer of culture and tradition (Eyerman & Turner, 1998).

Every generation that was born in a distinct era has its own unique traits. Born between 1965 and 1980, Generation X shares traits with earlier generations and desires to live under the same set of norms. Additionally, this generation enjoys learning new things and can keep up with technology advancements (Gross, 2017). The generation that is in the middle of working life and digital life is known as Generation Y, and they make up the fastest-growing segment of the labor force. They were born between 1981 and 1995. Millennials look for fulfilling careers that provide them the chance to grow and use their creative talents. They are technologically knowledgeable and use technology to improve the productivity of their work and leave a lasting impression on their employees (Twenge, 2018). The structure of Generation Z, those who were born between 1996 and 2010, allows them to think rapidly, adjust to change more quickly, engage in creative activities within a flexible framework, and keep up with the busy and fast-paced environment. They enjoy participating and producing answers from their point of view in addition to being collaborative. These people, who were raised in the digital age and in a technologically advanced world, are sociable. Instead of face-to-face interaction, their social skills are maintained through the interactive virtual world offered by the internet and digital platforms (Sladek & Grabinger, 2014).

When all generations are considered, it has been found that technological development typically contributes to the generational inequalities. With digitalization, new ideas have surfaced that highlight generational divides. The first of the new concepts, the digital nomad, is a generation that was born into technology, controls technology, and obtains all the information they require through social media. In contrast to the digital natives, who prefer to use technology as a tool, the digital nomad is a generation that can use technology as a tool and prefers to obtain information from libraries and books (Bayne & Ross, 2011).

Each generation has various business understandings, principles, and attitudes, which are also represented in business life (Clark, 2017). According to studies done on the various generations and their approaches to working life, the X generation makes up the majority of the executive group while the Y generation makes up the working group. Although the X generation has an authoritarian structure and a traditional management approach, which is inclined to work within a certain rule, the Y generation has begun to reject an authoritarian approach and more equality issues have come to the fore, which has pushed the Y generation to make breakthroughs in "doing their own business." Generation Z, which followed Generation Y, has distinct expectations for the workplace since they can create their own working and entrepreneurial settings and can get information more quickly thanks to social networks and technology (Kaplan & Çarıkçı, 2018).

GenZ in Business and Marketing (Workplace)

Being born during a time of modern digital technology is the key characteristic that sets Generation Z apart from previous generations. The expectations and ambitions for the business world will also differ for this generation, which uses digital methods and will be most powerful on social networks (Pichler et al., 2021). Because of their propensity for using technology, Generation Z is expected to change the way organizations operate and help enterprises thrive by positioning them differently. However, due to the distinct characteristics of GenZ, such as unlimited career understanding and weak organizational commitment, organizations should now switch from traditional human resources practices to a talent management approach in order to attract these individuals to the business, provide them with the necessary value, develop their careers, and keep them in the business (Taş et al., 2017).

Employers need to be prepared for GenZ, which is preparing to take its place in business life, because GenZ has different characteristics and behavior patterns than other generations (Priporas et al., 2017, 2020). For this reason, it is very important for managers to know how to manage this young and inexperienced generation, not how to manage them, but what kind of management style they will adopt for their behavior. Understanding the behaviors and business needs of GenZ will ensure success in companies (Schroth, 2019). With the changing time, expectations and values in business life also differ due to the behavioral differences between generations. GenZ, which has a different behavior structure from other generations, can also exhibit different approaches in business life. GenZ has different motivations and different needs than previous generations. It prioritizes the working environment to be more flexible, especially in business life, to advance different careers and jobs at the same time, to improve communication between people, and to feel comfortable in the working environment (Gaidhani et al., 2019). In addition, GenZ cares about cultural diversity and expects companies to take part in a more inclusive working environment and to have professional development opportunities (Agarwal & Vaghela, 2018). This generation, which is integrated with technology, attaches importance to the fact that the environment and working methods in which they work have a technological infrastructure in business life. In addition, GenZers think that the work to be done with technology will be more proactive and impressive (Gupta, 2020).

Marketing is an important building block where businesses market their products in a managerial way and make businesses successful. Marketing also differed greatly between generations (Roberts & Manolis, 2000). Especially with the increase in technology, the social media tools used in marketing have brought a new breath and developed new methods, especially for the Z generation, and developed with their contribution. When the research on marketing is examined, it has been observed that the influencers and interactive shares in the social media tools of the GenZ affect this generation as a consumer (Fromm & Read, 2018). Along with Digital Marketing, companies are trying to meet the needs of this generation by developing new strategies in line with the needs and demands of the surface generation. This generation, which will take its place in business life in the future, will develop new methods, especially with the development of technology, and will give direction to marketing. It is thought that this generation, who knows digital marketing closely, will make great contributions to businesses by analyzing the needs of their own generation more easily. Especially in marketing strategies, GenZ advocates that content that contains more egalitarian discourses and creates value should be at the forefront (Grow & Yang).

Talent management is one of the most important issues that GenZ seeks in business life and marketing-related studies. Talent management can be defined as the talent optimization required to hire the right employees at the right time in line with the needs of the business and to manage employees correctly with a good strategy (Schweyer, 2004). Generation Z prefers work environments where they can show their talents and develops their working principles in this direction. This generation, which will also work in a dynamic sector such as tourism, wants to be involved in a business process that has talent optimization and where they can feel comfortable in their working environment. It has been suggested that GenZers, who will work in the field of management and marketing in the tourism sector, want to be included in the business strategies developed with the mentioned talent management (Goh & Okumus, 2020).

Global Market and GenZ Needs: Digital Transformation

Generation Z has emerged as one of the most mysterious consumers since they are tech-savvy, digitally connected, and trained users of market technologies. According to 2019 statistics, GenZ accounted for 32% of the global population, or 2.47 billion of the 7.7 billion people in the world (Spitznagel, 2020). Generation Z customers have been characterized as the first generation of digital natives (Kahawandala & Peter, 2020). Compared to other generations, Millennials and Generation Z are more likely to be comfortable with the language of the digital world. As a result, interacting with technology on a daily basis from a young age influences the decision to become a digital native in this generation (Munsch, 2021). Generation Z has a focused outlook on sales and has little free time to decide whether something is worthwhile. Bypassing these market filters and grabbing GenZ's attention requires offering them enjoyable and immediately practical experiences (Finch, 2015). What is cognitively intriguing is Generation Z's ability to generate and implement social interests, make purchase decisions, and meet their requirements, as well as the behavior exhibited in their connections with technology, social media, and values (Singh, 2014). GenZ consumers are becoming more and more interested in cutting-edge products that offer technical solutions and allow them to use mobile devices and access the internet (Turner, 2015). One of the well-integrated into Generation Z's consumption strategy is virtual reality. The members of Generation Z have the highest level of interest in virtual reality as generation members. Generation Z members are willing to employ augmented and virtual reality in the purchasing environment since it provides many of the qualities they seek (Lógó, 2017). In comparison to earlier generations, GenZ values purchasing things with a brand name that makes them happy and values products that care about their image (Wansi, 2020). Currently, marketing plans for Generation Z customers are a requirement. The best approach for businesses to connect with them is to make intelligent, strategic, and purposeful use of platforms and technologies. According to Macura, there are eleven strategies to market GenZ (Macura, 2022):

- Creating channel-specific content such as LinkedIn, Snapchat, Instagram, Facebook, TV
- Making the content that is mobile-optimized and keeping it short
- Using videos a lot such as YouTube videos, TikTok videos
- Highlighting the people and principles that support the brand because GenZ prefers original brands
- Being transparent and accountable
- Positioning the influencer as the focal point of a strategy in and of itself
- Allowing Generation Z to participate in your marketing process
- Using platforms such as TikTok to stimulate content creation, engagement, and participation, and to include everyone in the creation process
- Content creation that is exciting and entertaining
- Taking advantage of user-generated content
- Continuing attempts with omni-channel marketing

Generation Z spends a substantial amount of time on social media sites and video source platforms such as Instagram, Snapchat, Facebook, and YouTube in pursuit of interesting and entertaining material (Nielsen, 2018). As a result, digital marketers and advertisers are being compelled to create focused campaigns in order to get the attention of these digital natives who will spend wisely on products and brands (Jiang & Stylos, 2021). According to this viewpoint, the digital transformation of corporate business structures and the changes detected in them are separated into three stages (Radzikhovska, 2021).

The first is *automation* which is the storage of business processes, data in its current form and in the enterprise's style of functioning, and the transfer of data to electronic systems for data exchange. The second is the *digitization* of the enhancement of particular business procedures through adherence to the instruments and innovations of the digital economy. The third is *digital transformation* which entails a qualitative shift in every aspect of an organization's business model, beginning with strategy and concluding with the digitalization of all management, core, and supporting business operations (OECD, 2017). Consequently, businesses who use digital transformation strategies to learn about Generation Z's preferences and what differentiate them from previous generations will be able to develop useful connections that benefit all parties involved.

Working Environment of GenZ

Employers are beginning to focus on how to meet the demands of Generation Z as they enter the twenty-first century. According to academics, each generational cohort has a distinct approach to work and the world of work (Harber, 2011). According to HR experts, Generation Z is innovative, caring, tolerant, socially responsible, emotional, knowledgeable, and individualistic (Racolţa-Paina & Irini, 2021). The characteristics, demands, traits, and work practices of this diverse, technology-enabled GenZ are still poorly understood. Therefore, employers and HR managers need to anticipate the requirements of the workplace and change their methods of recruiting, motivating, and retaining this new generation of young workers. Due to their connection to the digital world, Generation Z is also considered to be more imaginative, more entrepreneurial, and more open to commercial and global activities than previous generations (Willam, 2019). This is a generation that has higher expectations of the workplace than previous generations. Flexible work schedules, shared ideals, higher compensation, and more mental health support have become priorities for GenZ (Staglin, 2022). In addition, many GenZ employees are looking for a multidisciplinary and global focus in their work, with the assumption that this will provide them with opportunities for mobility and rich experiences (Moses & Garia, 2017). Schawbel describes five GenZ characteristics we should be aware of (Schawbel, 2014):

- There is evidence that they are becoming more entrepreneurial
- They report that money has less of an impact on them than it used to
- Traditional channels of communication are the ones they prefer

- Their desire for your respect and consideration
- Their desire for employment with a trustworthy boss

Transparency, confidence, flexibility, and personal freedom are undisputed GenZ preferences, and these are the qualities it expects from leaders. GenZ is looking for leaders who can motivate them in their day-to-day work, while at the same time inspiring them to do new things and develop their careers. Because technology plays an important role in their lives, they are looking for organizations that enable them to communicate and perform their professional duties without time restrictions. They also believe that education does not help solve real-world problems, so they prefer work environments that offer opportunities for professional development. GenZers are interested in ways to make the world a better place, especially ways to improve the lives of businesses and customers. When communicating with GenZ, managers need to set the right environment and tone, develop a business mentality that supports a mobile workforce, establish a corporate culture and create a welcoming work environment for GenZ, enable everyone to share information at any time, and encourage collaboration and sharing, and for GenZ employees, it is absolutely essential to interact and communicate meaningfully (Haycox, 2022). They want to work for companies that give back to the community and are transparent about the benefits of their location (Bredbenner, 2020). Understanding the demands of Generation Z can give leaders and managers a competitive edge, increase recruitment and retention rates, and, ultimately, build a better business (Baldonado, 2018). Therefore, in order for businesses to remain effective, it is important that they understand this new generation and adapt their cultural practices to match workplace ideals. By understanding their needs and demands and designing a workplace that meets these needs, organizations have a better chance of developing the loyalty of GenZ employees. Below are some actions that organizations can take to progress this initiative as they incorporate GenZ into their business lives as future employees (Andrews, 2022):

- A focus on the mental health and wellbeing of employees; the ability to offer excellent benefits and wages to help employees secure their financial future

- The provision of learning and skill-enrichment opportunities to enable employees to make the most of technology
- Providing free access to corporate information
- Highlighting diversity, equity, and inclusion; and never ignoring ethical considerations

This generation will want to work in an atmosphere that is as rewarding as the one it creates, which is what we should be doing now. Above are some of the critical issues that GenZ has brought to the workplace, which will also influence the future of organizations and how they're going to engage and retain them. Employers and managers will need to be prepared to accept the updated and digitally advanced employees of GenZ into the workplace.

GenZers as Marketing Managers in Tourism

Technological developments have an impact on all sectors of the economy and, as a result, on the way in which these sectors are managed. In the last few years, in particular, the pace of technological development has increased enormously, and this increase has made the world a more and more digital and global place (Keskin & Yazar, 2015). Undoubtedly, one of the sectors that is greatly affected by these changing and evolving trends is Tourism (Stylos, 2020). Being affected by both micro- and macro-level changes, the tourism sector has a sensitive structure. In tourism, as in any sector, technology has had a rapid impact on the dynamics of the sector. It is essential for businesses to adapt to this speed of technological change if they want to continue to exist and to benefit from this change. Otherwise, this speed of change can become a competitive disadvantage and lead to losing the traditional, familiar aspect of a business. This speed of technological development is creating different opportunities for the representatives of the different generations. In particular, Generation Z, who were born during the period when these technological developments were most prevalent and are more familiar with them than any other generation, can use them to their advantage (Stylos & Vassiliadis, 2023).

Understanding these new demands and standards is very important for a successful marketing strategy in a market where GenZ is an evolving market as an active consumer with new standards and demands (Thach et al., 2020). Therefore, a successful marketing manager should be able to have an understanding of today's changing standards and expectations and have the vision to meet them. In this respect, GenZ can be seen as having an advantage in understanding and responding creatively and proactively to their own generation, which is a significant part of the current and future consumer mass, as they are able to go beyond being mere consumers and find a relatively new position in management. As well as having these and other advantages, GenZers are confronted with a world that is increasingly digital, with a similar globalization, and with a social, economic, and political context that is strongly influenced at a young age by the Covid-19 epidemic (Nath, 2020). It is, therefore, necessary to confront a structure of uncertainty and various challenges, as well as of new possibilities and opportunities, where opportunities and risks intertwine (Entina et al., 2021). Given the highly vulnerable nature of the tourism sector to global and local crises, it is necessary to face these challenges, respond to changes, and develop rational solutions for all employees and generations in the tourism sector.

The digital world offers many marketing opportunities for instant contact with consumers and the diversification of contact channels. As a result, tourism organizations will be able to reach large audiences using innovative brand communication techniques. By using their skills and creativity to harness these opportunities for the tourism sector, GenZers can play a key role in breaking traditional patterns in the sector. Considering that they are more skilled than all other generations in reaching the necessary information in a quick and practical way (Matveeva & Krasnov, 2019), enabling the GenZ to use this skill will be an advantage in administrative processes.

In addition to their existing talents in the digital field, GenZ representatives need to be well equipped to be successful in managerial processes. Specifically, they need to be able to face the challenges, uncertainties, and complexities inherent in the tourism industry. This phenomenon of "volatility," "uncertainty," "complexity," and "uncertainty" that affects institutional environments is expressed as "VUCA" (Sarkar, 2016; Saleh &

Watson, 2017). Against the "VUCA" that also affects the tourism sector, GenZ members, especially in managerial positions, can increase their probability of being successful if they include their digital skills, avoid drowning in detail, focus on results, and think compactly.

Some of the positive features that distinguish GenZ from other generations are self-awareness, having a relatively more realistic view of life, innovative thinking, compact thinking, not drowning in details, and self-confidence (Tang et al., 2020). In addition, being able to react quickly to events is another positive differentiating feature of individuals in the GenZ (Agnes, 2020). These and similar positive features are the features that can create advantages and synergies at certain stages of the business life of the GenZ, but especially at the managerial levels, and that can be benefited by businesses in coping with the complexity of the VUCA world. Especially in the tourism sector, being able to react quickly to changes and thinking innovatively is very important in terms of sector dynamics. The VUCA world requires managers to have the skills to meet the requirements of the VUCA world. In this respect, features such as curiosity, creativity, and tolerance for complexity contribute to the development of the skill set suitable for the structure of this period. Studies show that the GenZ has more of these features compared to all other generations (Horstmeyer, 2019).

As more older generation managers retire or move up in their positions, there will be gaps in these positions in companies. GenZers are important candidates to fill these gaps because they have different skills and perspectives. Although GenZ is only just starting to find its place in new management levels, in the near future, many GenZers will have a say in these positions and will have the opportunity to reflect their views on the industry as they take on significant responsibility for the future of the industry. Therefore, it is important to make use of the skills of the GenZ and to create an environment where they can benefit from the experiences of the representatives of other generations at the point of shaping the sector of the future today.

As well as saying that business lines and ways of doing business will change, it would be realistic to predict that the tourism sector and other sectors will also reshape and change over time. As these changes change the sectors, they will also affect the business lines in the sectors, and business lines that may not exist today can be seen among the important career opportunities of the future. Barhate and Dirani (2021) share the prediction that perhaps as much as 90% of GenZ could work in occupations that do not currently exist. Based on this foresight, especially in the near future, the GenZ's openness to innovative thoughts and their ability to adapt quickly to changes can be counted among the features that can provide advantages both for the sector and for themselves. However, as in every generation, there are some negative distinctive features as well as positive distinguishing features for the GenZ. Generation Z lacks stable behavioral patterns and may be the most difficult generation to understand (Rodriguez et al., 2019), and individuals in this generation have a temporary desire to escape from the realities they face, as their attention span is more rapid and their enthusiasm can change rapidly (Wood, 2013). Nevertheless, the GenZ members, who can highlight the positive distinctive features they have and file the negative ones, will provide significant gains both for themselves and for the tourism sector.

Conclusion

Regarding both the tourism sector and other labor markets, it should be emphasized that GenZ has entered the labor market and companies need to prepare for them to enter different positions in the company. It is likely that by 2030, almost every entry-level job role in almost every industry will be filled by a member of GenZ (Al-Asfour & Lettau, 2014). Organizations are experiencing the coexistence of four different generations actively working in the workplace with GenZ, the newest entrants to the workforce. Businesses are witnessing the coexistence of four different generations actively working in the working environment (Barhate & Dirani, 2021). Organizations need to manage this diversity well in order for it to create wealth.

It is very important for companies to create an environment in which each generation can holistically deal with the advantages it offers over the dynamics of its own time, without making the mistake of proving the superiority of one generation over the other. As well as managing change, especially in a dynamic and highly open sector such as tourism, the preservation and transfer of past experiences in the context of corporate memory will ensure the sustainability of the continuity of the marketing success of companies. It should be emphasized that businesses that are able to create an environment that highlights the advantages of each of the X, Y, and Z generations and transfers these advantages to each other will be profitable, based on the literature (Lyons & Crown, 2014; Solaja & Ogunola, 2016; Barhate & Dirani, 2021). In particular, tourism businesses have the opportunity to embrace change in a multifaceted way by creating a rich corporate culture that develops the ability to turn these generational differences into an asset and to build bridges between different generations. It is possible in an environment where intergenerational disagreement is embraced, where knowledge is shared, and where different perspectives and input from the younger generation are encouraged by internal mentoring programs.

Although it is necessary to avoid drawing sharp boundaries and avoid absolute generalizations when talking about the characteristics of the generations, there is an idea about the distinctive features of the behaviors of adopting and using digital technologies, especially of the GenZ individuals, who have a more pragmatic desire for unique personal products and experiences compared to other generations. There seems to be unity (Topic & Mitchell, 2019). Compared to previous generations, GenZ has grown with an extremely complex media and technological environment, making them a knowledgeable social group in the use of technology and far more proficient than other generations. However, the GenZ also has some specific expectations from the workplace environment in which they work. At the beginning of these expectations are transparency, equality of opportunity, free expression, and a workplace away from hierarchy. Therefore, it is imperative that HR departments employing the GenZ are open to innovations and make their business processes fun and creative. In addition, tourism businesses that will give the GenZ opportunities in various business levels and especially in management should increase the flexibility of their institutional structures and reduce bureaucracy (Robinson & Schänzel, 2019). Even in the steps of ensuring this flexibility, basing business processes based on continuous learning and having fun while learning, transferring practice rather than theory, and reducing bureaucracy, GenZ members who have come to managerial positions will have important roles. Tourism businesses that can do this can embrace the opportunity to create change rather than manage it. They may also be able to adapt to the expectations of the new generations and become the businesses preferred by GenZ. Through these and similar endeavors, GenZ can make tourism a much more accessible, transparent, and inclusive place, using the opportunities created by digital.

Research shows that GenZ has a positive attitude toward working in the tourism industry and evaluates such career options as "exciting" and "adequate" (Goh & Lee, 2018). However, there is very little research that specifically addresses the position of the GenZ in working life, in line with the fact that the GenZ has just taken their place in the working life. Further efforts should be made to deepen the understanding of the role of GenZ in managerial processes in the tourism and other sectors. In this regard, future research should focus more on showing how the presence of GenZ in managerial positions changes the dynamics of the industry (Pichler et al., 2021). In addition, there is a need to take a closer look at the GenZ not only in terms of its differences from other generations, but especially in the context of the job market and the tourism sector.

Recommended Further Reading

Fromm, J., & Read, A. (2018). *Marketing to GenZ: The rules for reaching this vast--and very different--generation of influencers.*

Hu, M., Pantano, E., & Stylos, N. (2023). "Home alone" no more: How does the internet of things (IoT) enhance travellers' subjective well-being. *Technological Forecasting and Social Change, 192*. https://doi.org/10.1016/j.techfore.2023.122563

Jiang, Y., & Stylos, N. (2021). Triggers of consumers' enhanced digital engagement and the role of digital technologies in transforming the retail ecosystem during COVID-19 pandemic. *Technological Forecasting & Social Change, 172*. https://doi.org/10.1016/j.techfore.2021.121029

Koulopoulos, T., & Keldsen, D. (2016). *GenZ effect: The six forces shaping the future of business*. Routledge.

Maloni, M., Hiatt, M. S., & Campbell, S. (2019). Understanding the work values of GenZ business students. *The International Journal of Management Education, 17*(3), 100320.

Schroth, H. (2019). Are you ready for GenZ in the workplace? *California Management Review, 61*(3), 5–18.

Sladek, S., & Grabinger, A. (2014). GenZ. *Introducing the first generation of the twenty-first century Available at.*

Stylos, N. (2023). Intelligence augmentation via human-AI symbiosis: Formulating wise Systems for a Metasociety. In *Philosophy of artificial intelligence and its place in society* (pp. 301–314). IGI Global.

Stylos, N., Rahimi, R., Okumus, B., & Williams, S. (2021). *Generation Z marketing and management in tourism and hospitality.* Springer.

Stylos, N., & Zwiegelaar, J. (2019). *Big data as a game changer: How does it shape business intelligence within a tourism and hospitality industry context?* (pp. 163–181). Springer.

References

Agarwal, H., & Vaghela, P. (2018, December). Work values of GenZ: Bridging the gap to the next generation. In *National conference on innovative business management practices in 21st century* (Vol. 1, p. 26). Faculty of Management Studies, Parul University.

Agnes, U. (2020). Transformation of Z- generation in the context of globalization and place marketing: The case of Hungarian students. *Socio Economic Challenges, 4*(1), 28–35.

Al-Asfour, A., & Lettau, L. (2014). Strategies for leadership styles for multigenerational workforce. *Journal of Leadership, Accountability, and Ethics, 11*(2), 58–69.

Alwin, D. F., & McCammon, R. J. (2007). Rethinking generations. *Research in Human Development, 4*(3–4), 219–237. https://doi.org/10.1080/15427600701663072

Andrews, M. (2022). *Effectively communicating and working with your GenZ team members.* Retrieved January 25, 2023, from https://www.forbes.com/sites/forbesbusinesscouncil/2022/12/19/effectively-communicating-and-working-with-your-GenZ-team-members/?sh=719b6c141ba7

Arslan, A., & Staub, S. (2015). *Kuşak Teorisi ve İç Girişimcilik Üzerine Bir Araştırma.* Kafkas Üniversitesi İktisadi ve İdari Bilimler Fakültesi. 6/11, s.1–24. Retrieved January 25, 2023, from https://www.fastcompany.com/3049848/5-ways-the-workplace-needs-to-change-to-get-the-most-out-of-generation-z

Baldonado, A. M. (2018). Leadership and GenZ: Motivating GenZ workers and their impact to the future. *International Journal of Managerial Studies and Research (IJMSR), 6*(1), 56–60.

Barhate, B., & Dirani, K. M. (2021). Career aspirations of generation Z: A systematic literature review. *European Journal of Training and Development, 46*(1/2), 139–157.

Bayne, S., & Ross, J. (2011). 'Digital native' and 'digital immigrant' discourses: a critique. *Digital difference: Perspectives on online learning*, 159–169.

Bellou, V., Stylos, N., & Rahimi, R. (2018). Predicting hotel attractiveness via personality traits of applicants: The moderating role of self-esteem and work experience. *International Journal of Contemporary Hospitality Management, 30*(10), 3135–3155.

Bredbenner, J. (2020). Generation Z: A study of its workplace communication behaviors and future preferences (Doctoral dissertation, Wichita State University).

Clark, K. R. (2017). Managing multiple generations in the workplace. *Radiologic Technology, 88*(4), 379–396.

Entina, T., Karabulatova, I., Kormishova, A., Ekaterinovskaya, M., & Troyanskaya, M. (2021). Tourism industry Management in the Global Transformation: Meeting the needs of generation Z. *Polish Journal of Management Studies, 23*(2), 130–148.

Eyerman, R., & Turner, B. S. (1998). Outline of a theory of generations. *European Journal of Social Theory, 1*(1), 91–106. https://doi.org/10.1177/136843198001001007

Finch, J. (2015, May 4). *What is generation Z, and what does it want?* Retrieved January 20, 2023, from https://www.fastcompany.com/3045317/what-is-generation-z-and-what-does-it-want

Fromm, J., & Read, A. (2018). *Marketing to GenZ: The rules for reaching this vast--and very different--generation of influencers*. Amacom.

Gaidhani, S., Arora, L., & Sharma, B. K. (2019). Understanding the attitude of generation Z towards workplace. *International Journal of Management, Technology and Engineering, 9*(1), 2804–2812.

Goh, E., & Lee, C. (2018). A work force to be reckoned with: The emerging pivotal generation Z hospitality work force. *International Journal of Hospitality Management, 73*, 20–28.

Goh, E., & Okumus, F. (2020). Avoiding the hospitality workforce bubble: Strategies to attract and retain generation Z talent in the hospitality workforce. *Tourism Management Perspectives, 33*, 100603. https://doi.org/10.1016/j.tmp.2019.100603

Gross, T. S. (2017). *Rules of the millennial generation* (M. Tokyay, Trans.). Istanbul Kultur University Press, 136s.
Grow, J. M., & Yang, S. (2018). Generation-Z enters the advertising workplace: Expectations through a gendered lens. *Journal of Advertising Education, 22*(1), 7–22. https://doi.org/10.1177/1098048218768595
Gupta, R. (2020). How to incorporate technology in the workplace for GenZ employees. *European Journal of Molecular & Clinical Medicine, 7*(10), 3699–3707.
Harber, J. G. (2011). *Generations in the workplace: Similarities and differences.* Doctoral dissertation, East Tennessee State University.
Haycox, S. (2022). *5 tips for communicating with GenZ employees.* Retrieved January 25, 2023, from https://www.interactsoftware.com/blog/tips-for-communicating-with-GenZ-employees/
Horstmeyer, A. (2019). How VUCA is changing the learning landscape—And how curiosity can help. *Development and Learning in Organizations: An International Journal, 33*(1), 5–8.
Jiang, Y., & Stylos, N. (2021, November). Triggers of consumers' enhanced digital engagement and the role of digital technologies in transforming the retail ecosystem during COVID-19 pandemic. *Technological Forecasting & Social Change, 172*, 121029. https://doi.org/10.1016/j.techfore.2021.121029. Epub 2021 Jul 21. PMID: 36540888; PMCID: PMC9755634.
Joshi, A., Dencker, J. C., & Franz, G. (2011). Generations in organizations. *Research in Organizational Behavior, 31*, 177–205.
Kahawandala, N., & Peter, S. (2020, March). Factors affecting purchasing behavior of generation Z. In *Proceedings of the international conference on industrial engineering and operations management Dubai.*
Kaplan, B. T., & Çarıkçı, İ. H. (2018). İş dünyasında jenerasyonlar: X, Y ve Z jenerasyonları üzerine kavramsal bir inceleme. *Balkan ve Yakın Doğu Sosyal Bilimler Dergisi, 4*(1), 25–32.
Keskin, İ., & Yazar, T. (2015). Examining digital competence of teachers within the context of lifelong learning based on of the twenty-first century skills Öğretmenlerin yirmi birinci yüzyıl becerileri ışığında ve yaşam boyu öğrenme bağlamında dijital yeterliliklerinin incelenmesi. *Journal of new results in science, 12*, 1691–1711. https://doi.org/10.14687/IJHS.V12I2.3503
Lógó, E. (2017, September). Values derived from virtual reality shopping experience among generation Z. In *2017 8th IEEE international conference on cognitive infocommunications (CogInfoCom)* (pp. 000237–000242). IEEE.
Lyons, S., & Crown, L. (2014). Generational differences in the workplace: A review of the evidence and directions for future research. *Journal of Organizational Behavior, 35*(1), 139–157.

Mackay, H. (1997). Generations. *Independence, 22*(2), 15–17.

Macura, A. (2022). *Marketing to GenZ: How to do it the right way*. Retrieved January 20, 2023, from https://www.searchenginejournal.com/marketing-to-GenZ/471784/#close

Matveeva, A., & Krasnov, R. (2019). Generation "Z" as a generation for market of goods and services in digitization. In *International scientific and practical conference on digital economy (ISCDE 2019)* (pp. 609–612). Atlantis Press.

Moses, R., & Garia, N. (2017). *Generation Z enters the workforce Generational and technological challenges in entry-level jobs*. Retrieved January 25, 2023, from https://www2.deloitte.com/content/dam/insights/us/articles/4055_FoW-GenZ-entry-level-work/4055_FoW-GenZ-entry-level-work.pdf

Munsch, A. (2021). Millennial and generation Z digital marketing communication and advertising effectiveness: A qualitative exploration. *Journal of Global Scholars of Marketing Science, 31*(1), 10–29.

Nath, H. (2020). Covid-19: Macroeconomics impacts and policy issues in India. *Space and Culture, India, 8*(1), 1–13.

Nielsen. (2018). *Millennials on millennials: TV and digital news consumption*. Nielsen Insights.

OECD. (2017). *Meeting of the OECD council at ministerial level*. Retrieved January 20, 2023, from https://www.oecd.org/mcm/documents/C-MIN-2017-4%20EN.pdf

Pichler, S., Kohli, C., & Granitz, N. (2021). DITTO for GenZ: A framework for leveraging the uniqueness of the new generation. *Business Horizons, 64*(5), 599–610.

Priporas, C. V., Stylos, N., & Fotiadis, A. K. (2017). Generation Z consumers' expectations of interactions in smart retailing: A future agenda. *Computers in Human Behavior, 77*, 374–381.

Priporas, C. V., Stylos, N., & Kamenidou, I. E. (2020). City image, city brand personality and generation Z residents' life satisfaction under economic crisis: Predictors of city-related social media engagement. *Journal of Business Research, 119*, 453–463.

Racolţa-Paina, N. D., & Irini, R. D. (2021). Generation z in the workplace through the lenses of human resource professionals—A qualitative study. *CAL, 22*(183), 78–85.

Radzikhovska, Y. (2021). Digital transformation and its influence on changing the marketing orientation of business structures and consumer behavior. *Baltic Journal of Economic Studies, 7*(2), 200–209.

Roberts, J., & Manolis, C. (2000). Baby boomers and busters: An exploratory investigation of attitudes toward marketing, advertising and consumerism. *Journal of Consumer Marketing, 17*, 481–497. https://doi.org/10.1108/07363760010349911

Robinson, V., & Schänzel, H. (2019). A tourism inflex: Generation Z travel experiences. *Journal of Tourism Futures, 5*(2), 127–141.

Rodriguez, M., Boyer, S., Fleming, D., & Cohen, S. (2019). Managing the next generation of sales, GenZ/millennial cusp: An exploration of grit, entrepreneurship, and loyalty. *Journal of Business-to-Business Marketing, 26*(1), 43–55.

Saleh, A., & Watson, R. (2017). Business excellence in a volatile, uncertain, complex and ambiguous environment (BEVOCA). *The TQM Journal, 29*(5), 705–724.

Sarkar, A. (2016). We live in a VUCA world: The importance of responsible leadership. *Development and Learning in Organizations: An International Journal, 30*(3), 9–12.

Schawbel, D. (2014). *GenZ employees: The 5 attributes you need to know*. Retrieved January 25, 2023, from https://www.entrepreneur.com/growing-a-business/GenZ-employees-the-5-attributes-you-need-to-know/236560

Schroth, H. (2019). Are you ready for GenZ in the workplace? *California Management Review, 61*(3), 5–18.

Schweyer, A. (2004). *Are you managing your company's talent?* An interview with HR-IT consultant and author Allan Schweyer. Retrieved November 20, 2005, from http://www.optimizemag.com/article/showArticle.jhtml?articleId=18100229

Singh, A. (2014). Challenges and issues of generation Z. *IOSR Journal of Business and Management, 16*(7), 59–63.

Sladek, S., & Grabinger, A. (2014). *Gen Z: The first generation of the 21st Century has arrived!* Available at: https://www.xyzuniversity.com/research

Solaja, O., & Ogunola, A. (2016). Leadership style and multigenerational workforce: A call for work place agility in Nigerian public organizations. *International Journal of African and Asian Studies, 21*, 46–56.

Spitznagel, E. (2020). Generation Z is bigger than millennials — And they're out to change the world. *New York Post*. Retrieved January 20, 2023, from https://nypost.com/2020/01/25/generation-z-is-bigger-than-millennials-and-theyre-out-to-change-the-world/

Staglin, G. (2022). The future of work depends on supporting GenZ. Retrieved January 20, 2023, from https://www.forbes.com/sites/onemind/2022/07/22/the-future-of-work-depends-on-supporting-GenZ/?sh=64c8e499447a

Stylos, N. (2020). Technological evolution and tourist decision-making: a perspective article. *Tourism Review, 75*(1), 273–278. https://doi.org/10.1108/TR-05-2019-0167

Stylos, N., & Vassiliadis, C. A. (2023). Gamification design: Toward developing image perception scales for generation Z consumers. *International Journal of Contemporary Hospitality Management*. https://doi.org/10.1108/IJCHM-12-2022-1616

Tang, J., Tosun, C., & Baum, T. (2020). Do GenZs feel happy about their first job? A cultural values perspective from the hospitality and tourism industry. *International Journal of Contemporary Hospitality Management, 32*(12), 4017–4040.

Taş, H. Y., Demirdöğmez, M., & Küçükoğlu, M. (2017). Geleceğimiz olan Z kuşağının çalışma hayatına muhtemel etkileri. *Uluslararası Toplum Araştırmaları Dergisi, 7*(13), 1031–1048.

Thach, L., Riewe, S., & Camillo, A. (2020). Generational cohort theory and wine: Analyzing how GenZ differs from other American wine consuming generations. *International Journal of Wine Business Research*. https://doi.org/10.1108/ijwbr-12-2019-0061

Topic, M., & Mitchell, B. (2019). *Generation Z & consumers trends in the environment packaging*. Project report, Leeds, The Retail Institute.

Turner, A. (2015). Generation Z: Technology and social interest. *The Journal of Individual Psychology, 71*(2), 103–113.

Twenge, J. M. (2018). *İ- Nesli, Kaknüs Publications: 431s* (O. Gündüz, Trans., 1st ed.).

Wansi, J. (2020). *How do Instagram influencers affect the consumer buying behavior of GenZ*. Artevelde University of Applied Sciences.

Willam, D. (2019). *Millennials vs generation Z: What employers must know (infographic)*. Retrieved January 20, 2023, from https://smallbiztrends.com/2016/07/millennials-vs-generation-z.html

Williams, K. C., & Page, R. A. (2011). Marketing to the generations. *Journal of behavioral studies in business, 3*(1), 37–53.

Wood, S. (2013). *Generation Z as consumers: Trends and innovation*. Institute for Emerging Issues: NC State University.

10

Generation Z Perspective on Destination Image Restoration Through Dark Tourism

Chang (Cherise) Li and Elizabeth Agyeiwaah

Introduction

Marketing strategies to restore destination image for sustainable tourism development could be a way to minimize the damage of unexpected crises to destinations. During health-related crisis such as the COVID-19 pandemic, the negative health image of a destination can serve as a deterrent factor in the visitor's destination decision process which requires strategies that reverse such negative health perceptions into an attraction. Recent studies argue that tourists have less preference for destinations with more COVID-19-confirmed cases and long-haul leisure activities

C. Li (✉)
School of Business, Qingdao University, Qingdao, People's Republic of China
e-mail: lichang@qdu.edu.cn

E. Agyeiwaah
Faculty of Business, Economics and Law, The University of Queensland, Brisbane, Australia
e-mail: l.agyeiwaah@uq.edu.au; agyeiwaahelizabeth@yahoo.com

N. Stylos et al. (eds.), *Contemporary Marketing Management for Tourism and Hospitality*, https://doi.org/10.1007/978-3-031-65049-9_10

(Li et al., 2021), nonetheless, possible destination pull strategies to restore destination image are non-existent.

Short-haul and young tourists are seen as suitable targets for destination crisis recovery (Aro et al., 2009; Li et al., 2021). Generation Z as a young consumer group accounts for 32% of the global population and their influences and spending power are increasing (Lee & Wel, 2018), especially in the tourism and hotel industry (European Travel Commission, 2020; Stylos & Vassiliadis, 2023). At present, the existing research is not sufficient for the preference of Generation Z for different types of tourism, and the situation is even worse for dark tourism. From the perspective of generational differences, the development of leisure tourism also needs to grasp the influence of social and demographic trends on consumers behavior to provide products to better meet their needs and expectations (Haddouche & Salomone, 2018; Robinson & Schänzel, 2019). Consequently, this chapter aims to (1) examine young Chinese, especially Generation Z's perspectives on restoring unsafe destination image in an early period of health crisis; (2) formulate a framework for Generation Z's perceptions of destination image restoration; and (3) identify Generation Z's characteristics accordingly.

Embodied Cognition Theory and Tourism Destination Image (TDI)

As a multifaceted theoretical proposition, embodied cognition emphasizes that the cognitive process is related to the body and the external environment (Goldinger et al., 2016). Stimuli from the external environment on various aspects of the body affect an individual's perception; he/she understands the external world through a comprehensive grasp of bodily sensations and psychological states (Sun & Lv, 2021). Cognition represents perception, attention, symbolization, language, contextualization, association, imagination, memory, recall, logic, inference, and meaning (Selvam, 2018). These components of cognition emerge not only from the activities of the brain but also from the entire human body as it interacts with its environment (Colombetti, 2014; Johnson, 2021;

Selvam, 2018). Studies have shown that all aspects of cognition, are mediated, influenced, or even initiated by affect, which may even before human senses engage the environment in perception (Colombetti & Thompson, 2008; Johnson, 2021). In management research, the embodied cognition theory is a solid theoretical foundation of sensory marketing, even though its application in tourism remains scarce (Lv et al., 2019).

A growing stream of tourism research now recognizes the importance of embodied cognition since tourists' sensory experiences and bodily feelings provide a critical source of information that determines tourists' cognition, emotions, and behavior subconsciously (Kock & Ringberg, 2019). This importance has concomitantly driven research involving embodied cognition toward the perspective of tourists' experience, including experiential honeymoon, cultural creative design, sensory dimensions of tourist experiences, and dark tourism (Liu et al., 2019; Sun & Lv, 2021). However, from these preceding tourism studies, there is a lack of exploration of the marketing audience's perception. Particularly in the context of COVID-19, where tourism demand forecasting is a daunting task, one possible way is to understand the perspectives of potential domestic tourists by exploring their mental processes to propose strategies for TDI restoration (Polyzos et al., 2020).

Consequently, the mental activities that interact with the body and the environment that influence people's cognition, emotion, and behavior can also be identified in the composition of the TDI. TDI comprises cognitive image, affective image, and conative image, as elaborated below (Gartner, 1994; Michael et al., 2018; Woosnam et al., 2020):

- **Cognitive image** is generated based on tourists' understanding and evaluation of a destination. The evaluation aspects mainly include quality of experience, touristic attractions, environment and infrastructure, entertainment, activities, cultural traditions, etc.
- **Affective image** relates to the motives and feelings of a destination. For instance, pleasant, arousing, relaxing, and exciting are four indicators to measure this dimension.
- **Conative image** is analogous to behavior. It drives the way people act based on cognitive and affective images. For example, it can be identified from tourists' WOM and revisit intention.

Generally, the mixed cognitive and affective image elements are combined to identify tourism information consultants' role, who act as potential marketers and play a vital role in tourism promotion and TDI restoration (Garay Tamajón & Cànoves Valiente, 2017). This also leads to a problem that most of the existing studies focused on the cognitive-affective image and the relationship between them, but the conative image has been largely neglected (Woosnam et al., 2020). Significantly with the context of COVID-19, a deeper understanding will emerge if embodied cognition theory is combined with the three components of the TDI to restore the image of a destination for its potential tourists.

TDI Rebuilding After Health-Related Crisis

Destinations around the world have been suffering from varying degrees of image damage because of the COVID-19 pandemic, including but not limited to China, Malaysia, and India in the Asia-Pacific region, Italy and Spain in Europe, and Brazil in the Americas (Henry-Biabaud & Weltman, 2020). Similar to COVID-19, tourist destinations had faced the threats of different health-related crises, such as the outbreak of Ebola in Western Africa and the SARS spread in Asia (Jamal & Budke, 2020; Maphanga & Henama, 2019). For destinations, what comes with the health-related crisis is not only the loss of tourists' favor for an unpredictable period of time but also the destruction of TDI. Infectious diseases have been accompanied by the development of human civilization history and will coexist with humankind in the future, and such crises cannot be completely eliminated by medical means (McNeill & McNeill, 1998). Therefore, relevant tourism recovery strategies are needed both in the short term and in the long run. However, the current recovery strategies for health-related crisis response to the COVID-19 pandemic are limited and rarely focus on TDI. Some of them are general strategies and mainly focus on tourism policies and recovery strategies for certain types of the tourism industry (Hoang et al., 2020; Kreiner & Ram, 2020; Smart et al., 2021). Therefore, the elaboration of strategies to restore the image of destinations based on empirical evidence is of great significance to the post-crisis development of destinations.

A good TDI is a prerequisite for effective marketing strategies and for the destination to maintain general market competitiveness (Telisman-Kosuta, 1989). As a decisive component of destination tourism products, the TDI is also a key factor influencing consumers' travel decision-making (MacKay & Fesenmaier, 1997; Telisman-Kosuta, 1989). Based on the "dimorphic theory" (Gunn, 1972), Gartner (1994) identified eight sources of image formation, including

- Overt-induced image from institutions, through traditional forms of advertising;
- Overt induced from travel intermediaries, through requested information;
- Covert induced from celebrity endorsement, through recommendations;
- Covert induced from specialists from reports, stories, or articles;
- Autonomous from various actors through independent information such as reports, stories, or articles;
- Unsolicited organic from friends, through unsought information;
- Solicited organic from friends through WOM;
- Organic from personal travel experience.

From marketer and audience perspective, TDI could be classified into two types of images which are projected and perceived images (Andreu et al., 2000; Garay Tamajón & Cànoves Valiente, 2017; Marine-Roig & Ferrer-Rosell, 2018):

- **Projected image** is provided by the supply side. It is projected by destination management organizations or tourism marketers through marketing activities.
- **Perceived image** comes from the demand side. It is based on the organic image that is perceived by the tourists.

For tourism destination management, the projected image is more controllable compared to the perceived image (Garay Tamajón & Cànoves Valiente, 2017). Beerli and Martin (2004) explored the related dimensions and attributes that affect the formation of perceived TDI and found that safety is a key component that impacts TDI. Culture, history, and art also can help the destination enhance the perceived TDI.

One of the challenges TDI often faces is that the image perceived by tourists is usually inconsistent with the image projected by the supplier or destination management organization, and there may be significant differences between them (Marine-Roig & Ferrer-Rosell, 2018). For the construction of TDI, narrowing down the gaps between the perceived image and the projected image is important for destination promotion (Dinnie, 2008; Marine-Roig & Ferrer-Rosell, 2018). As the COVID-19 pandemic has tarnished the image of many countries due to the risk-averse nature of most tourists over infectious diseases, projecting a better image through practical strategies becomes crucial (Im et al., 2021). But the question is how. Thus, how can the gap between perceived image and projected image be corrected to reset tourism toward a better path? The connections between many tourist destinations with COVID-19 make them "unsafe" places in the minds of tourists threatening the destinations' attractiveness (Elizabeth et al., 2021; Karl et al., 2020). Yet, this perception of a lack of safety is connected to the death of COVID-19 patients, and such morbid identity could be transformed into a tourism product but only for adventurous and dark tourism lovers (Ariawan & Ahmat, 2020; Sharma et al., 2021). This chapter contributes to this perceived-projected image gap hypothesis by delving into a potentially adventurous group that represents active young consumers in China.

Dark Tourism and Generation Z

Dark tourism is seen as a niche market related to death. It satisfies individuals by providing an environment for their curiosity and obsession with death and drives people's commemoration of death and mortality without involving terror or fear (Light, 2017; Seaton, 2018). However, the term "dark tourism" is always closely related to blackness, death, and sadness in the association. The research on dark tourism has been given a "black" color from the very beginning, which started with the term "black spot" by Rojek (1993), after that the concept of dark travel was used (Foley & Lennon, 1996). In addition, other similar appellations in the literature include trauma tourism (Clark, 2006), morbid tourism (Blom, 2000), fright tourism (Bristow & Keenan, 2018), grief tourism (Lewis, 2008), thanatourism (Banaszkiewicz, 2015), and even death tourism (Ashby, 2017). The appellations connote sad experiences of the destinations, making people overlook the positive meaning it could bring to the destination.

Although dark tourism is attractive for tourists because of the past painful experiences of the destination (Martini & Buda, 2020), previous research has not taken into consideration the role of dark tourism in restoring the damaged TDI after crises or disasters. Current studies on dark tourism mainly focus on the conceptualization of dark tourism, supply and demand of dark tourism attractions, case studies of dark tourism, dark tourism in heritage sites, and tourists motivation for dark attractions (Light, 2017; Martini & Buda, 2020). Some scholars have also explored the range of dark tourism sites (Baldwin & Sharpley, 2009; Casella & Fennelly, 2016), and the development of dark tourism's typology (Dunkley et al., 2007; Sharpley, 2012).

From the tourists' perspective, some existing studies have demonstrated that dark tourism is, particularly, attractive to young tourists because they seek fear and excitement during travel to improve their level of enjoyment (Mura, 2010; Tinson et al., 2015). In China, the proportion of young tourists, especially those born after 1995 and known as Generation Z, have shown more interest in dark tourism than their other counterparts (Juodzbalis, 2018). Besides, dark tourism is an essential tourism type for youth groups, especially students, to be educated. This is consistent with the travel preference of Generation Z, who are willing to acquire knowledge through travel experience (OTA-Mafengwo, 2020; Wang et al., 2020; Priporas et al., 2017; Priporas et al., 2020). It is worth noting that an embodiment study with dark tourism sites in China found that sensory experience is crucial for this type of tourism (Xie & Sun, 2018). However, there is still a lack of empirical examination of the perceptions of the Chinese youth group on developing possible dark tourism strategies for Wuhan's tourism revitalization. Among the limited studies on young travelers and dark tourism, few works could be found in America and Asia (Bhati et al., 2020; Tinson et al., 2015). Scholars also point out that the geographical differences between tourists may lead to different reactions to dark tourism (Tinson et al., 2015). Consequently, the inclusion of Asian youth perspectives into the dark tourism literature will contribute to revitalizing the different narratives on dark tourism and enrich our understanding of this group.

Explore Generation Z's Perception of Destination Image Restoration

Grounded Theory Approach: Identify Generation Z and Y's Perception of TDI Restoration

During the research period, as the destination suffered the health crisis, Wuhan was stuck in an embarrassing period with rare tourists because of the early stage of unblocking. This chapter, therefore, targeted potential tourists as the research object instead. For the loss of tourism caused by COVID-19, the development of domestic tourists is a linchpin to compensate. Meanwhile, young consumers could be targeted for tourism recovery after COVID-19 (García et al., 2023; Sandiaga, 2021). Generation Z and Generation Y are the dominant compositions of the current young adult population. Among them, Generation Y, born after 1981, has been the subject of much research, while Generation Z, born after 1995, is still an incompletely understood group (OC&C Strategy Consultants, 2019). Comparing two adjacent generations is sometimes adopted by previous studies to identify and highlight Generation Z consumers' characteristics (Yazici & Arslan Ayazlar, 2021). Therefore, this chapter starts sampling with both Generation Y and Generation Z for data collection. On the one hand, this method helps us obtain as many strategies as possible for TDI restoration strategies of Wuhan. On the other hand, two sets of data from different generations help explore the salient characteristics of Generation Z consumers by comparison. This chapter employed the constructivist grounded theory approach (Charmaz, 2014). As a structured but flexible method, the grounded theory approach is suitable for exploring research fields with rare references and enables scholars to construct an explanatory theory to reveal the inherent process of the substantive research field (Chun Tie et al., 2019).

There are two stages of data collection. The results of the first round of focus group interviews provided specific directions for the second round of in-depth interviews. In the first stage, focus group interviews with 182 respondents who are Chinese Generation Z and Y were conducted via Zoom online meeting. It aimed to investigate respondents' perceptions of Wuhan's image and image restoration strategies in general. Respondents in the first-round interview indicated that they would like to visit Wuhan when it is a safe place after COVID-19 is gone. Before the interview, a video about the lockdown of Wuhan, comparison photos of Wuhan's scenic spots before and after COVID-19, and images of Wuhan being restricted from entry by other countries were broadcast to evoke deep emotions among respondents. After that, questions such as "What do you think of Wuhan before and after the COVID-19 crisis?"; "How did you perceive the impacts of the COVID-19 crisis for Wuhan to attract visitors?"; and "What will you suggest to change Wuhan's image?" were asked. The interaction between the focus group participants brought lively discussions and helped this chapter collect rich and meaningful data in the first stage (Gill & Baillie, 2018). Respondents' image perception of Wuhan before and during the COVID-19 outbreak was identified to confirm the existing gap between perceived and projected destination images. Five theme topics emerged from the first stage of data collection: the unsafe image of Wuhan; "Wuhan Spirit"; making Wuhan become a destination for young people; the possible positive effect of Wuhan after COVID-19; restoring Wuhan's destination image.

These themes guided the second stage of in-depth interviews for further strategies exploration with 39 respondents (29 Generation Z and 10 Generation Y). For example, for the topic "making Wuhan become a destination for young people," respondents were asked, "After COVID-19, how could Wuhan promote its destination image and attract young visitors?" For the topic "the possible positive effect of Wuhan after COVID-19", the interview questions included "After COVID-19, do you think Wuhan will be positively affected? Why?" and so on. With two stages of data collection, this chapter followed initial, focused, and theoretical coding approaches to arrive at the final conceptual framework.

This chapter ensured the reliability and trustworthiness of the data through suggested procedures in the literature (Chun Tie et al., 2019). First, the credibility of the data was ensured through a clear description and evidence of the transcribed text of the Wuhan destination image-restoring strategies proposed by China's Generation Z. In addition, at least two of the team members conducted frequent dialogues on the responses provided after each interview to ensure the dependability of the data. Cross-reviewing transcribed materials between authors to validate the emerging themes and triangulation were applied to ensure data transferability among team members. Confirmability is based on memo writing to record and reflect on the daily encounters of researchers and participants. Self-reflection and mutual supervision and participation allowed the authors to be cautious about their own biases, assumptions, and beliefs that may affect the research process (Cypress, 2017).

The final data of Wuhan's TDI restoration strategies were obtained from 39 respondents of China's Generation Z and Generation Y, the first respondent segment is 29 China's Generation Z (10 male and 19 female) aged 18–25 years. Their geographical distribution covered 15 provinces, municipalities, or special administrative regions in China. More than half of the respondents were females and one in three was males. The second segment of respondents includes 10 China's Generation Y (4 males and 6 females) aged 26–33, from nine provinces, municipalities, or special administrative regions. Three of the Generation Y respondents are students, and the rest of the Generation Y respondents are working in formal occupations. In comparison, the occupation of Generation Z respondents are mainly students and a few are young professionals, this finding is in line with the European Travel Commission's (2020) report on the occupational composition of adult Generation Z.

In regard to strategies for TDI restoration after the health crisis, Generation Z and Y share similar views in some aspects. Specifically, as typical products of special interest tourism, food, and cherry blossoms as Wuhan's tourism resources are highlighted by both Generation Z and Y consumers for TDI restoration. For instance, respondents propose to invite online celebrities to recommend food or restaurants, develop unique local food brands, and open up new food streets. Meanwhile, they expect a more distinctive experience than enjoying food and flower view only. That is why they propose the TDI restoration strategies through Han Costumes themed activity or cherry blossoms, as well as the combination of Han Costumes and cherry blossoms as special interest tour packages such as "*Wearing Han Costumes and Enjoying the Cherry Blossom*". Besides seeking uniqueness, it could be identified that both Generation Z and Y are very enthusiastic and respectful of Chinese traditional culture. The motivation for special interest tourism mainly comes from consumers' desire to participate in existing tourism and leisure activities in the destination or to develop new interests in a specific destination (Agarwal et al., 2018). Such type of tourism could enhance the image of the destination, enrich the experience of tourists, and benefit multi-party tourism product suppliers. In this condition, the development of special interest tourism is helpful for destinations to classify large markets into small market segments such as cultural tourism, heritage tourism, or event tourism. In a smaller and concentrated market segment, destinations could target specific tourists easily and provide special-interest travel packages to meet individual needs (Morrison, 2010). By participating in special interest tourism, such as flower viewing, food tourism, and costume tourism, tourists tend to have a deep and lasting memory of the destination and strengthen their connection with the destinations (Jin & Sparks, 2017).

Generational Characteristics and Dark Tourism Strategies for TDI Restoration

Through the analysis and comparison of the qualitative data collected from the two generations, this chapter recognizes that different from Generation Y, the Generation Z respondents raised up original series strategies for unsafe TDI restoration of destinations like Wuhan. They suggest that the COVID-19 health crisis could be used for dark tourism development, so as to enrich and transform the unsafe TDI of such destinations. The following section organizes and introduces the strategies proposed by Generation Z with the theme of dark tourism.

Table 10.1 summarizes and demonstrates the specific dark tourism strategies proposed by Generation Z. Referring to other dark destinations

Table 10.1 Dark tourism development and strategies to restore unsafe TDI

Related TDI components	Generated concepts	Strategies categorization	
Cognitive image	Anti-COVID-19 memorial hall	Memorial site development	Attractions development of dark tourism
Cognitive image	Recording names on the wall		
Cognitive image	Rebuilding a hospital	New dark tourism attractions development	
Cognitive image	Experiential museum		
Cognitive image	Theme park		
Affective image	Real-life heroes	People marketing	Marketing strategies development for dark tourism
Affective image	Patriotic attachment		
Affective image	Celebrity endorsement		
Affective image	Heroes' stories	Narratives	
Cognitive image	Museum collections		
Affective image	Film and television works		
Conative image	Traveling feedbacks	Word-of-mouth	
Affective image	Wuhan residents' voice		

Source: Authors

that also suffered painful experiences but are developed as a form of edutainment products for tourists to engage with past suffering experiences. Many Generation Z respondents, therefore, suggested that dark tourism development could apply to destination management to turn its unfortunate experiences into opportunities. As part of restoring Wuhan's TDI damaged by the COVID-19 pandemic, memorial sites and new dark tourism attractions are China's Generation Z expectation toward Wuhan's dark tourism development. Some Generation Z respondents suggest setting up a memorial site for remembrance, such as a memorial hall for anti-COVID-19 and recording names of heroes on the memorial wall. Their opinions disclose that pain and crisis also could become a legacy and have great potential to attract tourists (Ashworth & Hartmann, 2005). About one-third of Generation Z respondents suggested or agreed with the new dark tourism attraction development, such as building the hospital as a dark tourism attraction and using the hospital as a theme park or experiential museum of the COVID-19 pandemic.

Notably, more than half of Generation Z respondents emphasize visiting dark tourism attractions to commemorate the past while avoiding excessive seriousness and sorrow. This viewpoint highlights the ***light-hearted*** characteristics of China's Generation Z, that is, in the face of dark tourism products that are closely related to death and painful memories Generation Z is seeking technological interaction, entertainment, and education in an easier way. As one respondent proposed:

> *A COVID-19 pandemic museum or theme park can be reconstructed on the basis of the cabin hospital, in which visitors can experience the lives of patients or doctors during quarantine and medical treatment … I am interested in museums with high technology, VR technology could be applied for visitors to see the virus vividly and understand the process of its effect on the human body. Meanwhile, it could provide some interactive games, such as visitors pretending to be immune cells to fight against viruses.* (Hans, a 20-year-old male, from Guangdong)

Generation Z's preference is also in line with Wright (2018) that dark tourism may allow tourists to interact with past painful experiences in a form of entertainment. The tourists' learning enthusiasm is emphasized, dark tourism museums could become a type of attraction to satisfy Generation Z tourists by providing education and experience platforms. Some Generation Z respondents expressed their expectations for learning and gaining experience during visiting. In this chapter, the pleasant experience and interaction with new technology in dark tourism attractions are what many respondents yearn for, which is also in line with the traveling motivation of gaining educational experience and entertainment of tourists (Seraphin, 2017).

Applying people marketing, narratives, and WOM to dark tourism development are strategies proposed by Generation Z for TDI image restoration damaged by the pandemic crisis. People marketing is a new term generated in this chapter, which includes real-life heroes, patriotic attachment, and celebrity endorsement. The real-life hero is a new concept introduced by both Generation Z and Y respondents. It is similar to celebrities as an important element of enhancing destination brand identity (Alt, 2019). Different from celebrities, the "real-life heroes" are ordinary people inspired by patriotic attachments to a country or a region and made outstanding contributions during the crisis. These real-life heroes are praised by society, such as medical-related staff (e.g., doctors) and anti-pandemic people (e.g., couriers), soldiers, all Wuhan people, and frontline workers.

The "real-life heroes" and stories represent the "love for Wuhan" which could be seen as a kind of patriotic attachment during the crises. Promoting real-life hero stories through media and documentaries is a strategy proposed by both Generation Z and Y. Creatively, Generation Z respondents indicated that this strategy could also be combined with dark tourism. They suggest that real-life heroes could be helpful for TDI restoration as elements of memorial sites or new dark tourism attractions. Different from Generation Y, stronger admiration remarks related to people's contribution and attachment to society and the country are frequently affirmed and praised by many Generation Z respondents. From their statements related to real-life heroes who saved Wuhan people and the country Generation Z's ***patriotic*** characteristic is also identified. As one of the respondents stated:

> *During the COVID-19 outbreak, using heroes who had traveled thousands of miles to support Wuhan including decorators, drivers, carpenters, masons, and many female laborers is important. Some medical staff handed in a letter of "Request to the frontline" and built a high wall between us and the disease bravely. Sweat is dripping under their heavy protective clothing, but they were still focusing on patients' treatments day and night. They are true warriors and worthy of the respect of the entire society … After the COVID-19 pandemic, a museum or theme park related to the crisis can be built to record people and events in this outbreak.* (Hans, a 20-year-old male, from Guangdong)

Celebrity endorsement as a type of people marketing strategy is proposed by both Generation Z and Y respondents. This strategy is commonly adopted by destination branding (Baker & Legendre, 2019; Kumari, 2018). Based on one-third of Generation Z respondents' opinions, for the COVID-19 post-pandemic stage, celebrity endorsement is helpful to alleviate dangerous perceptions and prove the safe tourism environment, which could be applied to the dark tourism development of Wuhan. Different from Generation Y, Generation Z respondents' preferences show significant characteristics as ***digital natives*** that like short videos (e.g., TikTok) and admire internet celebrities (e.g., Key Opinion Leaders). Their suggestions include different forms of celebrity endorsement such as web celebrities' live shows of tourism attractions, travel bloggers' promotions, and celebrities' variety shows. For example:

> *The influence of celebrities may be greater ... the negative image of Wuhan related to insecurity caused by the virus can be reversed by influential people, such as famous stars ... if the celebrities are willing to go, fans will think this place may not be dangerous, and they will relieve their worries* (Star, a 20-year-old female, from Liaoning)

> *Instead of watching crisis-related TV documentaries that talk about being strong in the face of adversity, I believe key opinion leaders (KOL) live broadcasts are more attractive to young people like us ... Nowadays online KOLs are very popular; TikTok, Xiaohongshu (a Chinese social media and e-commerce platform), and other live broadcast platforms can also be used to attract the broad masses of youth audience.* (Cecilia, 19-year-old female, from Macau)

Narratives could be seen as an effective way of branding from the perspective of commercial entities (Zhong et al., 2017). Both Generation Z and Y respondents suggest using narratives to improve the negative destination image. Generation Y preferred daily reports on TV or Newspapers to declare the safe environment of Wuhan for TDI restoration. Generation Z proposed more diverse suggestions, from which three types of narratives emerged, namely, heroes' stories, museum collections, film, and television works. More than half of Generation Z respondents mentioned the narrative strategies to restore Wuhan's TDI with dark tourism development. They proposed specific ways for marketing promotion such as documenting hero characters and stories in the COVID-19 pandemic museum and recalling visitors' emotions and commemorating through the meaningful collections. For instance, two respondents state that:

> *COVID-19-related museums could exhibit collections that represent the living conditions of ordinary people during COVID-19 … and also should collect the graffiti walls of the hospital at that time, protective clothes with names written, and so on.* (Mike, a 20-year-old male, from Heilongjiang)

> *I think the red curtain is a representative collection to put in the COVID-19 museum. It is related to a series of well-known videos on TikTok, which is about a person in Wuhan who recorded a red curtain that had been floating outside a neighbor's window for several months because no one is at that house during the COVID-19 outbreak. In these videos, the tree outside the window has sprouted from the winter dead branches at the beginning of the outbreak and turned green, which means spring is coming. Many people were watching the videos and concerning about the house owner for his/her health. Finally, one day the owner came back … Lots of people were tracking this series of videos on TikTok, and they already have feelings attached to the red curtain.* (Cici, a 20-year-old female, from Sichuan)

Word-of-mouth is also proposed by Generation Z respondents as an effective strategy for dark tourism development and promotion. By emphasizing the communication purposes with unobvious commercial nature and wide usage in tourism products and services (Zhong et al., 2017), word-of-mouth is defined as "customers telling stories". In this chapter, respondents affirmed the positive role of word-of-mouth for destination promotion with the combination of Wuhan's dark tourism attractions. Generation Z respondents insisted that locals' voice and the tourists' after-visit feedback are key factors to improve their travel intentions. For example:

> *I think word-of-mouth from local university students will attract students from other places.* (Charlie, 21-year-old male, from Shanxi)

> *A package tour can be provided to tourists for free through an online platform lottery. The route design of this tour package should be more exquisite, and dark tourism museums or other attractions can be included. Travelers will be asked*

> *to write some travel diaries and then give feedback to the online platform. Because travelers went there as an amateur, their feedback will be more credible.* (Grace, a 20-year-old female, from Jiangxi)

From the strategies proposed by Generation Z, this chapter found they are ***cautious and distrustful*** (Robinson & Schänzel, 2019). They hope to confirm the safety of the destination in the voice of ordinary people through online love expression activity, live shows, non-professional travelers' feedbacks, friends, and locals' attitudes. The above preferences also illustrate that Generation Z respondents doubt official information channels such as TV, newspaper, or official media, information that is less official but intuitive is more real for them.

Embodied Cognition and Destination Image Restoration Framework

China's Generation Z, as an audience of perceived destination image and potential tourists, their brains and bodies interact with the environment affected by the crisis (Selvam, 2018; Xie & Sun, 2018). According to the embodied cognition theory, a person's cognition is caused by the brain and body's interaction with the environment (Colombetti, 2014; Johnson, 2021). Therefore, it is reasonable to understand that China's Generation Z is influenced by the environment with the COVID-19 pandemic background, which subconsciously affects their emotion, cognition, and behavior. This is the reason for suggesting dark tourism for Wuhan's TDI restoration, when they put forward specific strategies based on their own perception. These strategies could be categorized by emotion, cognition, and behavior (see Fig. 10.1). Their proposal is based on the cognition of the COVID-19 environment and the sad experience they perceived about Wuhan during the outbreak (Nong, 2020).

Since the damaged TDI of Wuhan was caused by the gap between the projected image and the perceived image, narrowing down this gap is the key to TDI restoration of this health crisis destination (Marine-Roig & Ferrer-Rosell, 2018). As shown in Fig. 10.1, the strategies provided by

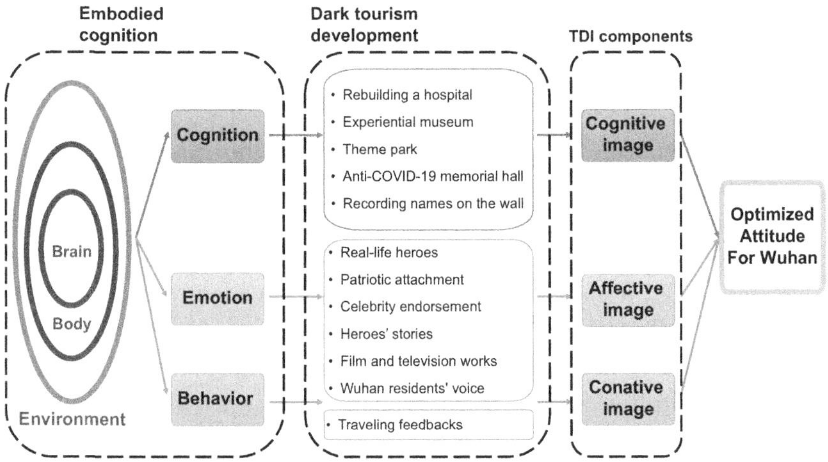

Fig. 10.1 TDI restoration framework. (Source: Authors)

China's Generation Z related to cognition, emotion, and behavior, which contribute to destination image components cognitive image, affective image, and conative image correspondingly. For instance, strategies such as rebuilding a hospital, experiential museum, theme park, anti-COVID-19 memorial hall, and recording names on the wall are related to attractions or infrastructure, entertainment, and activities within the dark tourism attractions. The above concepts will influence the cognitive image of the tourism destination (Hernández-Mogollón et al., 2018; Qu et al., 2011). Although real-life heroes, patriotic attachment, celebrity endorsement, heroes' stories, film and television works, and Wuhan residents' voices do not belong to the attractions of Wuhan, but with the combination of dark tourism development, those concepts affect people's feelings and emotions about Wuhan, which belongs to affective image category (Hernández-Mogollón et al., 2018; Qu et al., 2011). Traveling feedback is related to the conative image which is analogous to behavior after travel.

Characteristics of China's Generation Z Consumers

It is precisely because of the specific characteristics of this generation that they propose corresponding destination image restoration strategies. The characteristics of China's Generation Z are also significant for the follow-up tourism and leisure behaviors understanding about this generation. Figure 10.2 summarizes five characteristics of China's Generation Z based on the findings. The interpretation of study findings illustrates that Generation Z's strategies and perceptions are not only affected by the COVID-19 environment, but their unique cognition, emotion, and behavior are also related to their growing environment. Specifically, this cohort is ***digital natives*** and short video lovers (e.g., TikTok). They admire internet celebrities (e.g., KOLs). Their ***innovation and creativity*** (Corbisiero & Ruspini, 2018) are also closely related to their nature as digital natives. Creatively, they proposed COVID-19 experiential museums with technology and virtual reality instead of exhibition-style only. Meanwhile, they are ***cautious and distrustful*** (Robinson & Schänzel, 2019). They hope to confirm the safety of the destination in the voice of ordinary people through online love expression activity, live shows, non-professional travelers' feedback, and friends and locals' attitudes, which are more unofficial but intuitive and real for them. Compared with their

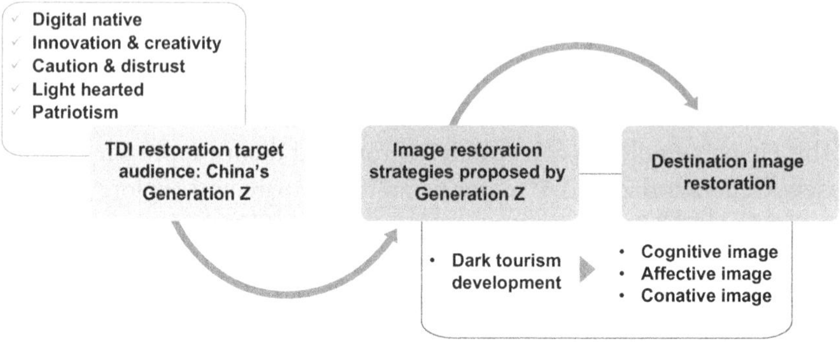

Fig. 10.2 Traits of Generation Z identified from proposed TDI strategies. (Source: Authors)

peers in other countries, China's Generation Z is closely and deeply aware of the rapid and uncontrollable spread of the virus, as well as the insignificance and helplessness of human beings. That might make them cautious about sources of information for destinations that have suffered the COVID-19 outbreak.

Notably, under the ***light-hearted*** group features, some Generation Z lamented that the positive stories (e.g., the red curtain story), pleasant experiences, and interaction with new technology within dark tourism attractions are what they yearn for. Goh and Okumus (2020) indicated the "fun factor" of Generation Z is that they regard work as a kind of fun. Here, the authors argue that Generation Z is not pure hedonists with the purpose of having fun; they are not showing disrespect for death and dark tourism in the COVID-19 context. But rather, they respect and realize the pain and sorrow that the health-related crisis has brought to mankind and prefer to commemorate the sad memories in a light-hearted form as a response mechanism (Corbisiero & Ruspini, 2018). Interestingly, the ***patriotic*** characteristic is specifically identified from China's Generation Z proposal and recognition of "real-life heroes" and the "patriotic attachment" under the heroism theme. They are not only recognizing the patriotism of ordinary people in COVID-19 but also showing their sensitivity and deep understanding of patriotism. This may be attributed to the construction of patriotism education having a profound impact on Generation Z. The Chinese government insists on starting with children and teenagers in patriotism education since 1994 (Ma & Liu, 2021).

Conclusion and Recommendations

This chapter employs a constructivist grounded theory approach to interview Generation Z and Y respondents to compare and explore the perspectives of China's Generation Z in unsafe TDI image restoration. Motivated by the embodied cognition theory and tourism destination image theoretical framework, this chapter proposes a qualitative model to explain strategies that can ignite the positive attitudes of Generation Z toward Wuhan with the negative image. The empirical findings point to strategies to develop dark tourism attractions which reflect embodied views of China's Generation Z. It provides a conceptual framework that highlights that China's Generation Z is influenced by the COVID-19 pandemic environment, which subconsciously affects their emotion, cognition, and behavior (Fig. 10.2). Given such findings, dark tourism strategies are generated for Wuhan's destination image restoration through Generation Z's perception. Through the grounded theory approach, the strategies of destination image restoration identified are not statistically tested. Hence, future studies may employ quantitative methods to test the proposed model and relationships, and a larger sample size with demographic variety is recommended.

Meanwhile, it should be noted that China's Generation Z is not only affected by the COVID-19 environment but also by their unique cognition, emotion, and behavior (Jiang & Stylos, 2021). Their characteristics are identified from the strategies they proposed (Fig. 10.2). Same with this cohort in other countries, China's Generation Z are digital natives with creativity, but to some extent, they are cautious and distrustful. In this chapter, their light-hearted and patriotic characteristics are also discovered. This chapter also demonstrates the possibility of exploring destination image restoration strategies from the perspective of marketing audiences and potential tourists and provides experience for follow-up research in terms of methods. This is also significant as a blueprint for future studies to extend research about China's Generation Z perception and behavior.

Recommended Further Reading

Avraham, E., & Ketter, E. (2017). Destination image repair while combatting crises: Tourism marketing in Africa. *Tourism Geographies, 19*(5), 780–800.

Crouch, D. (2000). Places around us: Embodied lay geographies in leisure and tourism. *Leisure Studies, 19*(2), 63–76.

İlhan, Ö. A., Balyalı, T. Ö., & Aktaş, S. G. (2022). Demographic change and operationalization of the landscape in tourism planning: Landscape perceptions of the generation Z. *Tourism Management Perspectives, 43*, 100988.

Kock, F., & Ringberg, T. (2019). Embodied cognition effects on tourist behavior. *Annals of Tourism Research, 78*(C).

Li, C., & Huang, X. (2022). How does COVID-19 risk perception affect wellness tourist intention: Findings on Chinese generation Z. *Sustainability, 15*(1), 141.
Marjerison, R. K., & Gan, S. (2020). Social media influencers' effect on Chinese gen Z consumers: Management and use of video content platforms. *Journal of Media Management and Entrepreneurship (JMME), 2*(2), 1–18.
Nechita, F., Demeter, R., Briciu, V.-A., Varelas, S., & Kavoura, A. (2019). Projected destination images versus visitor-generated visual content in Brasov, Transylvania. In *Strategic innovative marketing and tourism*. Springer.
Seemiller, C., & Grace, M. (2018). *Generation Z: A century in the making*. Routledge.
Stylos, N. (2023). Intelligence augmentation via human-AI Symbiosis: Formulating wise Systems for a Metasociety. In *Philosophy of artificial intelligence and its place in society* (pp. 301–314). IGI Global.
Stylos, N., & Zwiegelaar, J. (2019). *Big data as a game changer: How does it shape business intelligence within a tourism and hospitality industry context?* (pp. 163–181). Springer.
Thangavel, P., Pathak, P., & Chandra, B. (2022). Consumer decision-making style of gen Z: A generational cohort analysis. *Global Business Review, 23*(3), 710–728.
Wang, Z. (2008). National humiliation, history education, and the politics of historical memory: Patriotic education campaign in China. *International Studies Quarterly, 52*(4), 783–806.

References

Agarwal, S., Busby, G., & Huang, R. (2018). Special interest tourism: An introduction. In *Special interest tourism* (pp. 1–17). CABI.
Alt, M. (2019). Beyond Beethoven? Constructing the image of the German city Bonn through the sound art project Bonn Hoeren. *Sound Studies, 5*(2), 209–212.
Andreu, L., Bigne, J. E., & Cooper, C. (2000). Projected and perceived image of Spain as a tourist destination for British travellers. *Journal of Travel & Tourism Marketing, 9*(4), 47–67.
Ariawan, Z., & Ahmat, N. N. (2020). Expected tourist attractions after pandemic COVID-19. *International Journal of Human and Technology Interaction (IJHaTI), 4*(1), 107–112.

Aro, A. R., Vartti, A.-M., Schreck, M., Turtiainen, P., & Uutela, A. (2009). Willingness to take travel-related health risks—A study among Finnish tourists in Asia during the avian influenza outbreak. *International Journal of Behavioral Medicine, 16*(1), 68.

Ashby, M. A. (2017). *Death tourism: Disaster sites as recreational landscape.* Taylor & Francis.

Ashworth, G., & Hartmann, R. (2005). Introduction: Managing atrocity for tourism [in:] horror and human tragedy revisited. In G. Ashworth & R. Hartmann (Eds.), *The Management of Sites of atrocities for tourism* (pp. 7–10). Cognizant Communication Corporation.

Baker, M. A., & Legendre, T. S. (2019). *Celebrity endorsement, message framing, and online social support: The gateway bug to edible insect consumption.*

Baldwin, F., & Sharpley, R. (2009). Battlefield tourism: Bringing organised violence back to life. In *The darker side of travel* (pp. 186–206). Channel View Publications.

Banaszkiewicz, M. (2015). Tanatoturystyka: Doswiadczenie, pamiec, etyka/ Thanatourism: Experience, memory, ethics. *Politeja, 35*, 91.

Beerli, A., & Martin, J. D. (2004). Factors influencing destination image. *Annals of Tourism Research, 31*(3), 657–681.

Bhati, A., Agarwal, M., Tjayaindera, D. N., Aung, R., Thu, M., & Nguyen, T. M. T. (2020). Dark tourism in SouthEast Asia: Are young Asian travelers up for it? *International Journal of Hospitality & Tourism Administration*, 1–17.

Blom, T. (2000). Morbid tourism-a postmodern market niche with an example from Althorp. *Norsk Geografisk Tidsskrift, 54*(1), 29–36.

Bristow, R. S., & Keenan, D. (2018). A study of fright tourism during Hallowe'en. *E-review of Tourism Research, 15.*

Casella, E. C., & Fennelly, K. (2016). Ghosts of sorrow, sin and crime: Dark tourism and convict heritage in Van Diemen's land, Australia. *International Journal of Historical Archaeology, 20*(3), 506–520.

Charmaz, K. (2014). *Constructing grounded theory.* Sage.

Chun Tie, Y., Birks, M., & Francis, K. (2019). Grounded theory research: A design framework for novice researchers. *SAGE open medicine, 7*, 8.

Clark, L. B. (2006). Placed and displaced: Trauma memorials. In *Performance and place* (pp. 129–138). Springer.

Colombetti, G. (2014). *The feeling body: Affective science meets the enactive mind.* MIT press.

Colombetti, G., & Thompson, E. (2008). *The feeling body: Towards an enactive approach to emotion.*

Corbisiero, F., & Ruspini, E. (2018). Millennials and generation Z: Challenges and future perspectives for international tourism. *Journal of Tourism Futures, 4*(1), 3–104.

Cypress, B. S. (2017). Rigor or reliability and validity in qualitative research: Perspectives, strategies, reconceptualization, and recommendations. *Dimensions of Critical Care Nursing, 36*(4), 253–263.

Dinnie, K. (2008). Nation branding: Concepts, issues. *Practice, 2.*

Dunkley, R. A., Morgan, N., & Westwood, S. (2007). A shot in the dark? Developing a new conceptual framework for thanatourism. *Asian Journal of Tourism and Hospitality, 1*(1), 54–63.

Elizabeth, A., Adam, I., Dayour, F., & Badu Baiden, F. (2021). Perceived impacts of COVID-19 on risk perceptions, emotions, and travel intentions: Evidence from Macau higher educational institutions. *Tourism Recreation Research*, 1–17.

European Travel Commission. (2020). *Study on generation Z travellers.* Brussels. https://etc-corporate.org/uploads/2020/07/2020_ETC-Study-Generation-Z-Travellers.pdf

Foley, M., & Lennon, J. J. (1996). JFK and dark tourism: A fascination with assassination. *International Journal of Heritage Studies, 2*(4), 198–211.

Garay Tamajón, L., & Cànoves Valiente, G. (2017). Barcelona seen through the eyes of TripAdvisor: Actors, typologies and components of destination image in social media platforms. *Current Issues in Tourism, 20*(1), 33–37. https://doi.org/10.1080/13683500.2015.1073229

García, M. J., Chico, J. R., & Sánchez, A. R. P. (2023). Generation Z and intention to travel: Effects of the perception of COVID-19. *RIMCIS: Revista Internacional y Multidisciplinar en Ciencias Sociales, 12*(1), 40–68.

Gartner, W. C. (1994). Image formation process. *Journal of Travel & Tourism Marketing, 2*(2–3), 191–216.

Gill, P., & Baillie, J. (2018). Interviews and focus groups in qualitative research: An update for the digital age. *British Dental Journal, 225*(7), 668–672.

Goh, E., & Okumus, F. (2020). Avoiding the hospitality workforce bubble: Strategies to attract and retain generation Z talent in the hospitality workforce. *Tourism Management Perspectives, 33*, 100603.

Goldinger, S. D., Papesh, M. H., Barnhart, A. S., Hansen, W. A., & Hout, M. C. (2016). The poverty of embodied cognition. *Psychonomic Bulletin & Review, 23*(4), 959–978.

Gunn, C. (1972). *Vacationscape: Designing tourist regions.* Taylor and Francis.

Haddouche, H., & Salomone, C. (2018). Generation Z and the tourist experience: Tourist stories and use of social networks. *Journal of Tourism Futures.*

Henry-Biabaud, O., & Weltman, T. (2020). *Restoring post-COVID-19 APAC travel and destinations' E-reputation: What social sentiment data tell us*. https://www.hospitalitynet.org/video/69000456.html

Hernández-Mogollón, J. M., Duarte, P. A., & Folgado-Fernández, J. A. (2018). The contribution of cultural events to the formation of the cognitive and affective images of a tourist destination. *Journal of Destination Marketing & Management, 8*, 170–178.

Hoang, T. G., Truong, N. T., & Nguyen, T. M. (2020). The survival of hotels during the COVID-19 pandemic: A critical case study in Vietnam. *Service Business, 14*, 387–411. https://doi.org/10.1007/s11628-020-00421-w

Im, J., Kim, J., & Choeh, J. Y. (2021). COVID-19, social distancing, and risk-averse actions of hospitality and tourism consumers: A case of South Korea. *Journal of Destination Marketing & Management, 20*, 100566.

Jamal, T., & Budke, C. (2020). Tourism in a world with pandemics: Local-global responsibility and action. *Journal of Tourism Futures.*

Jiang, Y., & Stylos, N. (2021). Triggers of consumers' enhanced digital engagement and the role of digital technologies in transforming the retail ecosystem during COVID-19 pandemic. *Technological Forecasting and Social Change, 172*. https://doi.org/10.1016/j.techfore.2021.121029

Jin, X. C., & Sparks, B. (2017). Barriers to offering special interest tour products to the Chinese outbound group market. *Tourism Management, 59*, 205–215.

Johnson, M. (2021). *The meaning of the body*. University of Chicago Press.

Juodzbalis, A. (2018). *Factors determining Generation Z luxury goods consumer buying behavior in the China mainland and Hong Kong.* (PhD), Shanghai International Studies University, (IV).

Karl, M., Muskat, B., & Ritchie, B. W. (2020). Which travel risks are more salient for destination choice? An examination of the tourist's decision-making process. *Journal of Destination Marketing & Management, 18*, 100487.

Kock, F., & Ringberg, T. (2019). Embodied cognition effects on tourist behavior. *Annals of Tourism Research, 78*(C).

Kreiner, N. C., & Ram, Y. (2020). National tourism strategies during the Covid-19 pandemic. *Annals of Tourism Research.*

Kumari, P. (2018). Impact of celebrity endorsement on tourism industry in India. *International Journal of Research in Social Sciences, 8*(4), 952–963.

Lee, M., & Wel, L. (2018). Gen Z is set to outnumber millennials within a year. *Washington Post*. https://www.chicagotribune.com/nation-world/ct-gen-z-millennials-20180820-story.html

Lewis, C. (2008). Deconstructing grief tourism. *International Journal of the Humanities, 6*(6).
Li, X., Gong, J., Gao, B., & Yuan, P. (2021). Impacts of covid-19 on tourists' destination preferences: Evidence from China. *Annals of Tourism Research*, 103258.
Light, D. (2017). Progress in dark tourism and thanatourism research: An uneasy relationship with heritage tourism. *Tourism Management, 61*, 275–301.
Liu, Y., Wang, S., & Jie, X. (2019). *Experiential honeymoon tourism study from perspective of embodied cognition.* Paper presented at the 2019 5th international conference on information management (ICIM).
Lv, X., Xu, H., & Li, H. (2019). A review of tourism destination marketing from perspective of sensory marketing theory. *Tourism and Hospitality Prospects, 3*(4), 66.
Ma, J., & Liu, P. (2021, June 20). Hold high the spiritual banner of the Chinese communists-summary of national patriotism education demonstration base construction. *People's Daily.* http://qh.people.com.cn/n2/2021/0620/c182757-34784559.html
MacKay, K. J., & Fesenmaier, D. R. (1997). Pictorial element of destination in image formation. *Annals of Tourism Research, 24*(3), 537–565.
Maphanga, P. M., & Henama, U. S. (2019). The tourism impact of Ebola in Africa: Lessons on crisis management. *African Journal of Hospitality, Tourism and Leisure, 8*(3).
Marine-Roig, E., & Ferrer-Rosell, B. (2018). Measuring the gap between projected and perceived destination images of Catalonia using compositional analysis. *Tourism Management, 68*, 236–249.
Martini, A., & Buda, D. M. (2020). Dark tourism and affect: Framing places of death and disaster. *Current Issues in Tourism, 23*(6), 679–692.
McNeill, W. H., & McNeill, W. (1998). *Plagues and peoples.* Anchor.
Michael, N., James, R., & Michael, I. (2018). Australia's cognitive, affective and conative destination image: An Emirati tourist perspective. *Journal of Islamic Marketing.*
Morrison, A. (2010). *Hospitality and travel marketing* (4th International ed.). Delmar, Cengage Learning.
Mura, P. (2010). Scary… but I like it!' young tourists' perceptions of fear on holiday. *Journal of Tourism and Cultural Change, 8*(1–2), 30–49.
Nong, X. (2020). *Panoramic records of heroic city heroic people's feats, the large-scale documentary "Wuhan fight with COVID-19 Pandemic" shockingly debut.* https://baijiahao.baidu.com/s?id=1678768334636644441&wfr=spider&for=pc

OC&C Strategy Consultants. (2019). *A generation without borders.* https://www.occstrategy.com/media/2120/%E6%97%A0%E5%9B%BD%E7%95%8C%E7%9A%84%E4%B8%80%E4%BB%A3_online.pdf

OTA-Mafengwo. (2020). *"New Tourism" in the Post-COVID-19 Pandemic—Report on the Changes in Tourism Consumption of China's Generation Z.* http://www.mafengwo.cn/gonglve/zt-968.html

Polyzos, S., Samitas, A., & Spyridou, A. E. (2020). Tourism demand and the COVID-19 pandemic: An LSTM approach. *Tourism Recreation Research,* 1–13.

Priporas, C. V., Stylos, N., & Fotiadis, A. K. (2017). Generation Z consumers' expectations of interactions in smart retailing: A future agenda. *Computers in Human Behavior, 77*, 374–381.

Priporas, C. V., Stylos, N., & Kamenidou, I. E. (2020). City image, city brand personality and generation Z residents' life satisfaction under economic crisis: Predictors of city-related social media engagement. *Journal of Business Research, 119*, 453–463.

Qu, H., Kim, L. H., & Im, H. H. (2011). A model of destination branding: Integrating the concepts of the branding and destination image. *Tourism Management, 32*(3), 465–476.

Robinson, V. M., & Schänzel, H. A. (2019). A tourism inflex: Generation Z travel experiences. *Journal of Tourism Futures.*

Rojek, C. (1993). *Ways of escape: Modern transformations in leisure and travel.* Springer.

Sandiaga, S. U. (2021). *Destination insight series: Indonesia with PATA & BBC.*

Seaton, T. (2018). Dark tourism history. In *The Palgrave handbook of dark tourism studies* (pp. 1–2).

Selvam, R. (2018). The science of embodied cognition and enactive emotion: Implications for improving outcomes in all therapies. https://integralsomaticpsychology.com/science-embodied-cognition-and-enactive-emotion-implications-for-improving-outcomes-in-all-therapies/

Seraphin, H. (2017). Terrorism and tourism in France: The limitations of dark tourism. *Worldwide Hospitality and Tourism Themes.*

Sharma, G. D., Thomas, A., & Paul, J. (2021). Reviving tourism industry post-COVID-19: A resilience-based framework. *Tourism Management Perspectives, 37*, 100786.

Sharpley, R. (2012). Towards an understanding of 'genocide tourism': An analysis of visitors' accounts of their experience of recent genocide sites. In *Contemporary tourist experience: Concepts and consequences* (pp. 95–109).

Smart, K., Ma, E., Qu, H., & Ding, L. (2021). COVID-19 impacts, coping strategies, and management reflection: A lodging industry case. *International Journal of Hospitality Management, 94*, 102859.

Stylos, N., & Vassiliadis, C. A. (2023). Gamification design: Toward developing image perception scales for generation Z consumers. *International Journal of Contemporary Hospitality Management.* https://doi.org/10.1108/IJCHM-12-2022-1616

Sun, J., & Lv, X. (2021). Feeling dark, seeing dark: Mind–body in dark tourism. *Annals of Tourism Research, 86*, 103087.

Telisman-Kosuta, N. (1989). Tourist destination image. *Tourist destination image.*, 557–561.

Tinson, J. S., Saren, M. A., & Roth, B. E. (2015). Exploring the role of dark tourism in the creation of national identity of young Americans. *Journal of Marketing Management, 31*(7–8), 856–880.

Wang, J., Wang, G., & Wang, X. (2020). Dark tourism research: Critical reflection and theoretical construction. *Tourism and Hospitality Prospects, 4*(6), 22.

Woosnam, K. M., Stylidis, D., & Ivkov, M. (2020). Explaining conative destination image through cognitive and affective destination image and emotional solidarity with residents. *Journal of Sustainable Tourism, 28*(6), 917–935.

Wright, D. W. M. (2018). Terror park: A future theme park in 2100. *Futures, 96*, 1–22.

Xie, Y., & Sun, J. (2018). How does embodiment work in dark tourism “field”? Based on visitors’ experience in memorial hall of the victims in Nanjing massacre. *International Journal of Tourism Cities.*

Yazici, S., & Arslan Ayazlar, R. (2021). Generation Z and tourism entrepreneurship: Generation Z’s career path in the tourism industry. In *Generation Z marketing and Management in Tourism and Hospitality* (pp. 53–99). Springer.

Zhong, Y. Y. S., Busser, J., & Baloglu, S. (2017). A model of memorable tourism experience: The effects on satisfaction, affective commitment, and storytelling. *Tourism Analysis, 22*(2), 201–217.

11

Responsible Tourism as a Vehicle to Achieving the SDGs

Alison Maton

Introduction

This chapter aims to analyse how responsible tourism marketing and management can aid in achieving the United Nations Sustainable Development Goals (SDGs), through the adoption of responsible destination marketing management and marketing approaches.

Adopting a case study approach, the chapter offers meaningful insights into the critical role the SDGs play to employ the principles of contemporary responsible tourism marketing and management. Therefore, reducing the potential for tourism to affect irreparable environmental, economic and socio-cultural damage to destinations and utilising tourism as a force for good to empower its communities, create economic stability and protect its natural environments.

A. Maton (✉)
Buckinghamshire New University, High Wycombe, UK
e-mail: alison.maton@bnu.ac.uk

N. Stylos et al. (eds.), *Contemporary Marketing Management for Tourism and Hospitality*, https://doi.org/10.1007/978-3-031-65049-9_11

Although tourism has the potential to contribute either directly or indirectly to all of the SDGs, this chapter will focus on the specific Goals which the UNWTO (2023a) has tourism specifically included as targets:

- Goal 8: Decent Work and Economic Growth, which focusses upon inclusive and sustainable economic growth,
- Goal 12: Responsible Consumption and Production, focussing upon sustainable consumption and production (SCP) and
- Goal 14: Life Below Water, focussing upon the sustainable use of oceans and marine resources.

The chapter also considers Goal 5: Gender Equality and Empower all Women and Girls, due to the significant contribution they make to the sector's labour force. There is a significant gender divide in the industry, with a reliance on labour in the tourism sector being provided by women and girls. Overall, 54% of roles in the global tourism industry are fulfilled by women (WTTC, 2021), however, they also account for 45% of global employment losses (United Nations, 2022), illustrating the precarious nature of their employment status.

Finally, it will also explore the potential link that adopting the SDGs as part of responsible tourism marketing and management has upon increasing destination resilience (Beirman, 2022).

Defining Responsible Tourism

The term responsible tourism is often used interchangeably with related concepts including sustainable tourism and eco-tourism, but in practice, it is distinctly different. Holloway and Humphreys (2022) reason that advocates of sustainability have long argued that responsible tourism should be defined as underpinning a properly thought-out management strategy, which harnesses the power and expertise of collaboration between the private and public sectors to prevent irreparable damage to a destination. This is further supported by Goodwin (2011) and Lade et al. (2020), who both define responsible tourism as the "vehicle" through which sustainable development is pursued.

If sustainability is the goal, then responsible tourism is how to get there. It is about the decisions and actions people take to make tourism more sustainable in the long run. (Goodwin, 2011)

This chapter will begin by outlining the Millennium Development Goals (MDGs) and their development into the Sustainable Development Goals (SDGs) within the context of responsible tourism.

From the Millennium Development Goals to the Sustainable Development Goals

In 2000, United Nations member states unanimously adopted the Millennium Declaration, which led to the development of the MDGs. These were a set of 15 pledges to be achieved by the year 2015, with the overarching aim to reduce extreme poverty (United Nations, 2023a).

Building on these, the SDGs are a set of 17 global goals with 169 targets, set by the United Nations which underpin the 2030 Agenda for Sustainable Development, depicted in Figs. 11.1 and 11.2.

The overarching aim of the SDGs is to focus on health for all, end poverty, protect the planet's resources and ensure that everyone can enjoy and benefit from peace and prosperity, now and for generations to come. Currently, 193 UN member states have pledged to adopt the SDGs and strive to do all they can to meet them. Observance of the SDGs is designed to ensure that the tourism industry avoids being perceived as one that socially and economically exploits its workers, by commoditising the community and using them as a source of cheap labour (Beirman, 2022).

Fig. 11.1 United Nations sustainable development goals 2015. (Source: United Nations, Public Domain, 2023a)

Responsible Tourism Marketing and Management as a Vehicle to Achieving the SDGs

The selected SDGs will now be covered in more detail, noting their links to the tourism industry and potential to employ Responsible Tourism Marketing and Management techniques to contribute towards their achievement.

Goal 8: Decent Work and Economic Growth

Decent work opportunities in the tourism industry can be notoriously difficult to secure, especially in destinations where there is a lack of planning and a dependency upon a high-volume, low-value approach to tourism. However, according to the OECD (2020), it has long been recognised that tourism is one of the largest, fastest and most consistent growth sectors in the world's economy. Its ability to generate jobs, export revenue

Fig. 11.2 Tourism for the SDGs. (Source: UNWTO, Public Domain, 2022a)

and domestic value-added make destinations not only more attractive to visit but to live, work and invest too. They note that, as such, tourism can make considerable contributions to socio-economic growth by providing diverse employment opportunities, benefits for urban and regional economies, opportunities for innovative and small-scale business operations, support for promotion of cultural authenticity and natural assets and support for infrastructure-related development and maintenance. Despite this encouraging outlook, there is still a clear gap between the higher level of quality and security of employment opportunities offered to men, as opposed to that afforded to women (International Labour Organization, 2022). Potential ways to address this are by introducing policies that encourage improved diversification through tourism value chains, reducing leakage, which, in turn, will increase the positive socio-economic impacts of tourism. As the tourism industry makes up 30% of the world's trade in services and has multiple links to other economic sectors, it can generate development opportunities across the entire tourism value chain (UNWTO, 2023a).

One destination achieving this is Bhutan. The Bhutanese government has long focused on measuring the gross national happiness of its people, as opposed to measuring its GDP as a marker of success. Moreover, they have long adopted a low-volume, high-value approach to tourism planning and development, both of which are reflected in the vision statement of their tourism policy:

> *To promote Bhutan as an exclusive travel destination based on Gross National Happiness (GNH) Values.* (Bhutan Department of Tourism, 2021)

Their focus upon sustainability at the heart of their numerous tourism policies over the years ensures that not only a high-quality tourism product is able to be delivered for the long-term but also it provides sustainable and profitable working conditions for its people.

One way they achieve this is by enforcing a tourism levy, known as the "Sustainable Development Fee". As stated in the Kingdom's Tourist Rules and Regulations, each tourist is liable to pay a levy of USD 200 per night, with a preferential fee for visitors from India charged at Nu 1200 (USD 15) per person, per night (Bhutan Travel, 2023). This money is invested in programmes that will directly benefit all Bhutanese, especially their young people. These initiatives focus upon the key themes of sustaining cultural traditions, protecting the environment, upgrading the country's infrastructure and building their resilience. In practice, this includes building schools, maintaining forests and wildlife, renovating their historical Dzongs and providing free healthcare.

The levy has seen a significant increase post-pandemic from USD 65 to USD 200 per person, per night. This could price out of the market a significant number of travellers who identify themselves as responsible tourists, and instead attract only those who can afford to travel here. This is indeed a technique to maintain a high-value, low-volume approach to tourism, but if used in isolation, it will not mean that their desired type of tourist is visiting, which could result in a significant amount of this levy being spent on remedial work. The levy increase did successfully limit the number of inbound tourists to Bhutan. However, such was its impact at reducing visitor numbers that earlier this year Bhutan introduced incentives for travellers. This included allowing travellers to increase their length of stay by an additional four nights exempted from the levy and the ability to purchase duty-free gold (Bhutia, 2023a, 2023b). Increasing length of stay will help exacerbate some of the positive impacts of tourism, such as increasing both primary and secondary expenditure in the destination and mitigate some of the negative ones such as reducing the need for repeat visits, thus reducing air-travel related carbon emissions, thus it aligns with their focus on sustainability.

From a broader perspective, there are concerns about how these tourism-related taxes are spent and what influence they may have on demand, as well as to what extent the tourists as tax payers are consulted in how it is spent (Cetin et al., 2017). Conversely, Suresh et al. (2022) found that tourists are willing to pay significantly higher tourist taxes for nature conservation in the case of Sri Lanka, especially elephant conservation. Furthermore, Borges et al. (2020) found in their evaluation of municipal tourist tax awareness in Porto, that just over half of those included in the study were aware of the tourist tax, and that this depended upon their socio-demographic characteristics and almost 71% thought the tax was "acceptable". Therefore, the impact upon demand a tourist levy will impose is dependent upon multiple factors, including the socio-demographics of the tourist-generating markets, level of demand and factors driving demand to the destination, the scale of the levy and tourist consultation regarding tax expenditure. This, in particular, will be challenging, as it would require a designated body to devise a method in which to consult tourists and gain consensus as to the "best" use of the levy.

Goal 12: Responsible Consumption and Production

Due to the climate emergency, there is an increased urgency for the tourism sector to accelerate its shift towards sustainability by adopting sustainable consumption and production modes. According to the UNWTO (2023a), tools to monitor sustainable development impacts for tourism, including for energy, water, waste, biodiversity and job creation will result in enhanced economic, social and environmental outcomes. Therefore, it stands to reason that Goal 12 can be utilised in conjunction with Goal 8, as adopting sustainable consumption and production (SCP) models can have a positive impact on job creation.

Mashpi Lodge in Ecuador is a luxurious lodge set in 2500 acres of cloud forest and rainforest where guests can encounter 400 species of birds, frogs, trees and endemic species that are not found anywhere else in the world, as depicted in Fig. 11.3. It achieves all of this whilst ensuring its operation has the minimum impact on its surroundings by employing myriad sustainable consumption measures. Moreover, they empower their local communities by employing local people to work at the lodge, and as employees, they become shareholders in the business. They also offer professional training and English classes to enable their progression (Mashpi Lodge, 2023).

Bordeaux is another destination that has recently positioned itself to appeal to responsible tourists. The planning and marketing work by the Bordeaux Tourism and Conventions Office has resulted in it being designated the European Capital of Smart Tourism and one of the world's most liveable cities. This has included participating in the Global Destination Sustainability Index since 2019, gaining certifications including Qualité Tourisme, Tourism & Handicap, Accueil Vélo (bike-friendly), and ISO 20121 on the responsible management of its event activity, and the "Innovative and Sustainable Destination" label (Bordeaux Tourism, 2023).

As illustrated in Fig. 11.4, Bordeaux's 8 Clear Commitments to Responsible Tourism, align with Goal 12: Responsible Consumption and Production. There is a focus on reducing waste, reducing the city's carbon footprint and preserving natural resources.

Fig. 11.3 Expedition trails surrounding Mashpi Lodge. (Source: Wiki Commons, Public Domain (1991))

Goal 14: Life Below Water

According to the UNTWO (2023), coastal and maritime tourism rely upon healthy marine ecosystems. Tourism development must be a part of an Integrated Coastal Zone Management so it can help preserve delicate marine ecosystems and promote a blue economy and the sustainable use of marine resources.

Since 2011, Mexico has had a significant problem with a macroalgae called sargassum, which is spoiling its beaches and coastlines from Cancún to Tulum. Normally, this would become a significant issue in the peak summer months; however, due to unprecedented weather events, its arrival has started as early as February 2023. In 2018, there was a particularly notable bloom of Sargassum, which caused a decline in occupancy rates at hotels at Riviera Maya by 2.87%. The hotels had to offer compensation to customers which included reduced prices, tours of off-beach attractions and moving them to other unaffected properties. Due to its

Bordeaux Tourist and Congress Office's 8 Clear Commitments to Responsible Tourism

1. Reducing Waste.
2. Reducing the City's carbon footprint and preserving natural resources.
3. Strengthen the promotion of an alternative tourist offer based on culture and nature activities in the metropolitan area, raising awareness of the preservation of their environment.
4. Promoting the professional integration of people living in a long way from job opportunities. Developing accessibility to people with disabilities or in 'precarious situations'.
5. Acting for a peaceful relationship between professionals, inhabitants and visitors. Ensuring that all live in harmony.
6. Promoting sectors of excellence and the consumption of local products.
7. Improving the Bordeaux Tourist and Congress Office's performance on sustainability.
8. Increasing the sustainable skills of local tourism professionals.

Bordeaux Tourism and Conventions, (2023).

Fig. 11.4 Bordeaux tourist and congress office's 8 clear commitments to responsible tourism. (Source: Bordeaux Tourism and Conventions (2023))

unpleasant odour and unsightly appearance, resorts are employing full-time staff dedicated to continually removing it before it can spoil the tourist experience. The clean-up and disposal operation costs hotels in the region of $200,000. This is not a sustainable approach, and it is not addressing the root cause or the problem (National Geographic, 2021).

Goal 5: Gender Equality and Empower All Women and Girls

According to the UNWTO (2022b), "*Tourism can be a tool for women to become fully engaged and lead in every aspect of society*". This can be viewed not only as women playing a major part in the supply offside of the tourism industry, by the generation of employment associated with this sector but also their engagement with the demand side of it through accessible tourism products (see Fig. 11.5).

Fig. 11.5 Gender equality and empower women and girls. (Source: Wiki Commons, Creative Commons Zero (2022))

Research conducted by the WTTC in 2018 found that in both developed and developing economies, the tourism sector employs a far higher proportion of women than other sectors, with 60% of the labour force in hotels globally made up by women. Not only are more women employed in the tourism industry than men, but crucially to the achievement of Goal 5, they have more opportunities for advancement (WTTC, 2018). Their research also found that in Bulgaria women occupy 71% of management and administrative roles in the sector compared with just 29% across all sectors in the country. Also, in Indonesia, Malaysia, Thailand and the Philippines, more than half of tourism businesses are run by women. With the figures being similar across Latin America.

The effect of the COVID-19 pandemic resulted in 62 million jobs being lost in 2020, leaving just 271 million employed across the sector globally. This was a decrease of 18.6% across the entire sector, with small and medium-sized enterprises being disproportionately particularly affected, alongside women, the young and minorities (WTTC, 2022). However, tour operators such as Responsible Travel are leading the way with this by offering products such as women only tours in Morocco. Such tourism products contribute to achieving Goal 5: Gender equality and empowering all women and girls.

As demonstrated in the examples above, although the SDGs provide a solid foundation for planning, their implementation in the realms of responsible tourism can be challenging. One of the key challenges of implementation is identifying who should take responsibility for responsible tourism.

Responsible Tourism—Whose Responsibility Is It Anyway?

According to Goodwin (2011), responsible tourism requires action on the part of the producers, tourists and all stakeholders to make decisions and demonstrate action on sustainable solutions from the originating markets to the destination. Lade et al. (2020) consider this to be a systems approach to responsible tourism and that adopting this way of thinking is that it focuses on how elements of the system interact with each other over time, as opposed to analysing individual elements on their own. They further argue that the future of responsible tourism requires decision-makers to adopt a systems approach, and that the interconnectedness and complexity of the tourism system is accounted for, and stakeholders play an active role in each part of the system.

Ensuring everyone takes their share of the responsibility is a fair and equitable approach to employing responsible tourism principles and practices. However, this is easier said than done. Lack of education, the gap between academia and practice and myriad external pressures, none more so than the current cost of living crisis sweeping across Europe and the continued recovery of the industry post-pandemic. Furthermore, there is considerable research published on the attitudes of tourists and their behaviour differing to that of when they are in their home environment, as they view their holidays as a break from everyday life (Becken, 2004; Dolnicar & Grün, 2009). This attitude alone could result in a considerable, if unintentional, raft of negative environmental and socio-cultural impacts. Depending upon whether they behave responsibly or irresponsibly, tourists' behaviour impacts sustainable development to varying degrees (Becken et al., 2003; Metz et al., 2007).

As global tourism is expected to continue to recover following the pandemic and myriad economic challenges in many nations, the challenge here remains whose responsibility is it to enact responsible tourism anyway. With destination marketing and management budgets being cut and structures remodelled (as is the case in England following the de Bois review), and the demand for travel remaining high, the challenge is not about limiting growth, but about managing it in a way where it can benefit all destination stakeholders and the environment. It is likely that there will be no one universal way to achieve this, although strategic tourism planning should form the foundations (Edgell, 2016; Edgell & Swanson, 2019), whilst adopting a destination-specific approach. This approach is demonstrated in the end of chapter case study, Greenland's Destination Management Strategy and the SDGs.

Strategic destination planning is widely accepted as being applicable to most tourist destinations, as it is "*action-orientated and concerned with handling unexpected events, if used within a framework of integrated long-range policy and planning*" (Edgell & Swanson, 2019, p. 243). However, this approach is not unfailing, as can be demonstrated in the instance of the rapidly developing tourism economy of Albania. They have recently reported record tourism arrivals of 7,543,817 visitors in 2022, an increase from 3,672,591 in 2014 (Instat, 2023). Despite having a National Strategy for Tourism Development, which is focused around sustainable tourism (Republic of Albania Ministry of Tourism and Environment, 2018), Albania is currently attracting large amounts of young hedonistic tourists from the United Kingdom and Europe due to their range of music festivals. This niche of tourism is oft cited as problematic due to its frequent associations with myriad negative touristic impacts, including all types of pollution, increase in crime, appropriation of the culture, alienation of the resident populations and environmental degradation (Pinke-Sziva et al., 2019). With their natural environment featuring as a key part of their tourism product and strategy, the implementation and operationalisation of this plan may benefit from revision. Following the lead of destinations such as Greenland and West Sweden would help them to refocus. West Sweden's current tourism strategy "Stepping up Sustainability" is focused around four key principles:

1. As little unnecessary environmental impact as possible;
2. Good for both residents and visitors;
3. More visitors when and where places are not full;
4. More full-time jobs and more resilient businesses (West Sweden Tourist Board, 2023).

However, they already have the well-developed infrastructure of eco-hotels, a well-developed public transport network and eco-attractions such as kayaking tours in place, areas that a developing country such as Albania would need to improve. This would therefore be a long-term approach for this country, but given the right conditions and guidance, one that would be beneficial to the sustainability of their tourism industry and environment.

Opportunities and Risks of the SDGs and Responsible Tourism

In theory, the SDGs have been widely accepted by the global tourism industry, via the leadership of the UNWTO. They have been incorporated as policy by all key transnational, national and sectoral tourism body industries (Beirman, 2022). However, there is considerable evidence to suggest that some tourism businesses and destinations ignore these goals. This could be particularly attributed to those poorer, developing destinations that are more likely to follow the economic model of tourism development, defined by Hall (2008), as the use of tourism to create employment, earn foreign revenue and improve terms of trade, encourage regional development and overcome regional economic disparities.

An example of this is Thailand. Thailand's tourism industry is significant to the economy, with the latest data available showing that it contributes 6.9% to GDP (UNWTO, 2018). As with other destinations, the country experienced a steep decline in international tourist arrivals during this time. To keep as much of the industry operating as possible, they switched their attention to focus on domestic tourism and long-term stays, which would lessen the impact of the then-mandatory quarantine period (IMF, 2021).

Melissen and Sauer (2019) note that the SDGs encourage innovation, rather than being viewed as something that must be done based on meeting specific requirements. They consider that goals such as Sustainable Consumption and Production (SDG 12) could become a driving force for new business models. This can be seen in the aforementioned the Women Only tours offered by Responsible Travel and in numerous other initiatives such as the Portuguese Government's 10-year public-private strategy, which aligns with Goal 12 and has been built with key players in the industry, who all share the overarching aim for Portugal to become the leading sustainable tourism destination in the world.

The SDGs were set on a global scale, thus providing a clear reference point for partnerships and other initiatives. However, they are not legally binding, they're a political agreement, setting national targets for countries, rather than for companies (Melissen & Sauer, 2019). This creates a significant challenge for the countries trying to persuade the companies which operate within it to conduct their business in a manner which aligns with the UN Sustainability Principles. Furthermore, Melissen and Sauer (2019) reason that companies could easily ignore the SDGs and refer to national governments as the ones who bear the responsibility.

Conversely, companies could utilise the SDGs as an opportunity to bolster their Corporate Social Responsibility policies and reports to increase their attractiveness and garner higher levels of public favour, as there are no accreditation bodies or measurement policies in place to determine who qualifies for assisting in achieving these goals and who does not. Beirman (2022) further notes that many transnational tourism associations (e.g. UNWTO, WTTC, PATA) annually award and recognise the destinations and tourism enterprises which fulfil all or most of the SDGs. Thus, "sustainability sells" as many businesses, both those directly and indirectly involved in the tourism industry, have recognised that apart from the positive ethical associations with sustainable tourism, there is also a growing sector of the tourism market that is attracted to sustainable destinations, experiences and products, which has considerable economic gain. However, this could lead to the SDGs becoming a target for greenwashing (Melissen & Sauer, 2019), thus removing itself from the core principles of responsible tourism.

Moreover, the tourism industry is notoriously susceptible to external shocks including economic downturns, wars, and the omnipresent threat of terrorism. Moreover, the industry's ongoing recovery from the COVID-19 pandemic, with the IMF (2020) musing that tourism-dependent countries are likely to feel the negative impacts of this crisis for much longer than other economies.

The tourism industry has shown encouraging signs of growth since it was all but forced to close in 2020 due to the outbreak and spread of COVID-19, with the UNWTO reporting that international tourist arrivals have increased by 65% between March 2022 and March 2023 (UNWTO, 2023b). The pandemic forced all tourism businesses to re-evaluate the traditional high-volume, low-value model of doing business (Beirman, 2022). However, as much as this was true, the destinations that are most reliant on tourism for their contributions to GDP were the ones that reopened their borders to international tourists the quickest, for example, Thailand.

The "Visit Thailand Year 2022" campaign was primarily focused on economic recovery, using tourism as its vehicle. During this year, Thailand welcomed 11.8 million international tourists and recorded 189 million domestic trips (Tourism Authority of Thailand, 2022a). Thus, reflecting that this multi-stakeholder campaign has led to a successful recovery of the tourism industry.

However, the campaign Thailand's Tourism Authority are now focusing on is "*responsible and sustainable tourism*" which was debuted at World Travel Market (WTM) London in 2022. At their WTM stand, they only featured tourism companies whose business approach aligns with sustainable principles. Moreover, they focused on tourism products and services that reflected the triple-bottom-line approach to sustainable tourism. This included tourism products that promoted environmental sustainability, and the inclusion of local communities, arts and culture from cities around Thailand that potentially matched the growing trend and travel behaviour of the UK and European markets, particularly targeting these markets in their cold winter months to enjoy the temperate climate in Thailand at this time of year. They also showcased 20 low-carbon destinations in Thailand, secondary tourism destinations and regenerative tourism initiatives in line with the Royal Thai Government's Bio-Circular-Green or BCG Economy Model (Tourism Authority of Thailand, 2022b).

Outside of the tourism industry and academia, the level of knowledge about responsible tourism products remains disparate. The current cost of living crisis affecting the United Kingdom has currently had minimal impact on the outbound and domestic tourism industries. Research by ABTA (2022) has shown that consumers are reticent to give up their leisure travel and are increasingly actively seeking out responsible tourism products. However, a study by the Word Economic Forum (2022) notes that whilst travellers do intend to travel more sustainably, there are still significant gaps between consumers' intentions and their behaviour. Currently, tourists are seeking the best possible value for money, with ABTA predicting there will be high demand for all-inclusive holidays to help people balance their budgets in 2023 (ABTA, 2022). This may present a significant barrier to engagement with responsible tourism products, as there is a perceived price premium associated with sustainable travel products compared to their traditional counterparts. Furthermore, the World Economic Forum (2022) notes that the willingness to pay this premium will vary by consumer segment, and the higher price causes customers to opt for more affordable alternatives, with those with less disposable income being disproportionately affected.

Adopting the SDGs as Part of Responsible Tourism Marketing and Management and Increased Destination Resilience

Beirman (2022) posits resilience in the realm of tourism applies to enterprises, sectors and destinations. Hall (2018) expand upon this further by noting that resilience involves the ability of an individual enterprise, sector or destination to recover rapidly from negative events and circumstances to the extent that the reputation, viability and marketability are compromised or lost during the crisis are rapidly restored. This does not just concern addressing the problems that have led to the crisis but also repairing the reputation, trust and confidence of the destination or enterprise's stakeholders, and thus their viability (Beirman, 2022).

These negative events or circumstances oft felt by destinations can be explained by applying Holling's 1986 Adaptive Cycle (Holling, 1986 in Gunderson et al., 2012), as it explains how the "sudden shocks" have various effects on a destination which may affect the destination's resilience (Liburd & Edwards, 2018). The changes that have occurred as a result of the shock may strengthen the destination or make it weaker. In this case, destination managers learn from the event and aim to rebuild it through approaches such as supplier collaboration and innovation (Hjalager, 2018).

One such shock was the COVID-19 pandemic. In an attempt to curtail the spread of the virus, countries were forced to close their borders at short notice for prolonged periods of time. Whilst this provided temporary relief to destinations from environmental degradation (Beirman, 2022), it also led to a raft of global significant socio-economic challenges, including mass unemployment, economic crisis and psychological instability (Bae & Chang, 2020). Unlike other shocks and crises that adversely affect the tourism industry, such as hurricanes, floods, forest fires and terrorism (Dolnicar & Zare, 2020), pandemics can cause long-term damage as they are likely to reoccur after the initial outbreak is under control (Üngüren & Güçlü, 2021). Adopting a strategic approach to tourism planning can help to lessen the shocks and provide a "road map" to recovery when the worst happens. Edgell and Swanson (2019) note that strategic tourism planning is a framework designed to provide direction for a tourism destination at any level: local, regional and national.

A fitting example of this framework in practice is Greenland. Their current destination management strategy, Towards More Tourism—A Joint Effort for Growth, applies the principles of strategic tourism planning and has aligned their strategy with four of the SDGs.

Case Study: Greenland's Destination Management Strategy and the SDGs

In 2016, the Government of Greenland created a strategy called "What It Takes". This focussed on creating more favourable conditions for tourism development. This was done through increasing cruise fees, planning new international runways, establishing landing infrastructure for cruise ships, promoting private investments in the experience and the start-up of visitor centres, market development via Visit Greenland and business initiatives (Visit Greenland, 2021).

Towards More Tourism—A Joint Effort for Growth, Innovation and Responsible Tourism, is Visit Greenland's current marketing and market development strategy, which focuses on further developing their tourism industry. The vision of this strategy is to "*help create economic growth, jobs and the sustainable development of our community by making Greenland an internationally known and sought after adventure destination with a focus on quality, safety and sustainability*" (Visit Greenland, 2021).

Their main focus is on attracting Adventure Tourists, as they state this is "*a tourism form that is rapidly growing internationally and that offers the most value for the local community, minimizes environmental impact and is considerate of local culture*" (Visit Greenland, 2021, p. 8). Throughout Towards More Tourism, sustainability and responsible tourism are golden threads that run throughout. They aim to work cooperatively with destination stakeholders to ensure that tourism is developed sustainably there, with it making positive environmental, economic and social contributions for both the local population and tourists.

Their goals (referred to as "Must Wins") reflect the importance of building a sustainable tourism industry, with a focus on year-round tourism (to increase job security and average length of stay and spend), and promoting favourable framework conditions to ensure that tourism develops in a way that fits the Greenlandic way of life, rather than tourism influencing and changing their culture and landscape. There are significant plans to improve access by 2024, with three new airports being developed and creating landing infrastructure for cruise ships. These developments to their infrastructure signal that they are totally committed to developing their tourism industry. Although they have clearly stated that the goals must be fulfilled within the life of the strategy by 2024, they will also act as a benchmark, for the long-term sustainable development and management of the destination.

Greenland's 4 "Must Wins"

1. Increase demand from adventure tourists.
2. All year-round tourism in Greenland.
3. Knowledge sharing and competence upgrade.
4. Promote favourable framework conditions.

This sustainable approach to tourism development is further evidenced by their strategy aligning with four of the SDGs, that is, SDGs 4, 8, 12 and 17, and they have pledged to fulfil these four SDGs by the end of this strategy in 2024. For each of the SDGs that they've selected to align with, they have chosen specific sub-goals, which effectively underpin their "Must Wins". This is a robust approach and providing they're successful in meeting their "Must Wins", then they should fulfil their commitment to meeting the aforementioned SDGs.

They have also referred to studies of tourism development from their Nordic neighbours, particularly The Faroe Islands and Iceland. Both destinations have experienced a rapid increase in tourists in recent years, and as a result, they have significantly developed their tourist infrastructure and access. 'The success of their Icelandic neighbours', however, has led to a range of scholars suggesting that they may be suffering from Overtourism (Dodds & Butler, 2019; Milano et al., 2019), a moniker that Greenland would be keen to avoid. With high satisfaction scores of 4.7 out of 5 being given by tourists for their overall experience and at least 80% of the population agreeing that tourism contributes positively to their society, this robust strategy is likely to deliver on its goals and meet the SDGs whilst doing so.

Conclusions

Through a range of case studies, this chapter has analysed how responsible tourism can be used as a vehicle to make positive contributions in achieving the United Nations Sustainable Development Goals (SDGs), via the adoption of responsible destination planning, management and marketing approaches. This has been done by examining in detail the three SDGs that have tourism specifically included as a target: SDGs 8, 12 and 14. SDG 5 was also included due to the significant contribution women and girls make to the industry's workforce.

Whilst there are significant steps being made in the right direction by destinations and tourism organisations to meet all of the SDGs, there is still a seismic amount of work still to be done. Speaking at a recent United Nations convention, UN Women Executive Director Sima Bahous stated that at the current rate of progress, SDG 5 would not be achieved for another 300 years. Due to the substantial contribution women make to the tourism industry globally, if SDG 5 is off track, then it stands to reason that the SDGs cannot be achieved by 2030 without gender equality (United Nations, 2023a).

The responsibility of responsible tourism lies with us all, but as was explored in this chapter, this is easier said than done. Myriad challenges such as the perceived price premium, a lack of education for tourists about responsible tourism (and whose responsibility is it that they receive this education), the gap between academia and practice, tourists' differing attitudes whilst on holiday, the continued high demand for all-inclusive package travel and the range of considerable external shocks that the world is currently facing make this an ongoing challenge. However, it is a challenge that is worth perusing, as there is evidence to suggest that adopting the SDGs as part of responsible tourism marketing and management can increase destination resilience, something that almost all governments and destination planners should be concerned with for the future sustainability of their destinations.

The progress towards meeting the SDGs has certainly been impacted by the pandemic. The urgency to reopen borders to tourists as soon as was safe to do so was of the utmost importance for developing countries who are dependent upon an economic model of tourism development has had a detrimental impact on their progress. It is evident that destinations with an overdependency on tourism for their economic welfare should seek to rebalance their economies, but this would not be a quick fix and would take a considerable attitudinal shift from not only key decision makers in their governments but also from tourists whom many have become accustomed to having "access" to the world. However, as was seen in the Greenland case study, this is achievable, although they do have the privilege of being economically stable and learning lessons from the more developed tourism industries of their neighbouring countries.

The SDGs provide a clear, widely recognised framework that can be easily adopted by destination planners to aid their efforts towards responsible marketing and management of their destinations. As such, they should be more widely incorporated into future destination strategies at national, regional and local levels.

Recommended Further Reading

Bhutan Department of Tourism. (2021). *Tourism policy.* [Online]. https://www.tourism.gov.bt/about-us/tourism-policy

dos Santos, R. A., Méxas, M. P., Meirino, M. J., Sampaio, M. C., & Costa, H. G. (2020). Criteria for assessing a sustainable hotel business. *Journal of Cleaner Production, 262*, 121347.

Edgell, D. L., & Swanson, J. R. (2019). *Tourism policy and planning: Yesterday, today and tomorrow* (3rd ed.). Routledge.

Goodwin, H. (2011). *Taking responsibility for tourism.* Goodfellow Publishers.

Goodwin, H. (2016). *Responsible tourism: Using tourism for sustainable development* (2nd ed.). Goodfellow.

Goodwin, H. (2022). www.responsibletourismpartnership.org

Hjalager, A.-M. (2018). Suppliers as key collaborators for sustainable tourism. In J. Liburd & D. Edwards (Eds.), *Collaboration for sustainable tourism development.* Goodfellow Publishers.

Spenceley, A. (2012). *Responsible tourism: Critical issues for conservation and development.* Routledge.

Stylos, N., Koroneos, C., Roset, J., González-Sánchez, C., Xydis, G., & Muñoz, F. S. (2018). Exergy as an indicator for enhancing evaluation of environmen-

tal management performance in the hospitality industry. *Journal of Cleaner Production, 198*, 1503–1514.
Stylos, N., & Vassiliadis, C. (2015). Differences in sustainable management between four-and five-star hotels regarding the perceptions of three-pillar sustainability. *Journal of Hospitality Marketing & Management, 24*(8), 791–825.
Responsible Travel. (2023). *Responsible travel website.* [Online]. Retrieved February 26, 2023, from https://www.responsibletravel.com/
TAT. (2022). *Proof of urgency to reopen borders for tourism.* Retrieved January 19, 2023, from https://www.thailanda.ro/wp-content/uploads/2022/06/Newsletter-June-2022.pdf
United Nations. (2023a). *The 17 Goals.* [Online]. Retrieved February 26, 2023, from https://sdgs.un.org/goals
WTTC. (2021). *The Cancun women's initiative declaration supporting women in travel & tourism: Global declaration and platform: Women's equality and leadership.* [Online]. Retrieved February 9, 2023, from https://wttc.org/Portals/0/Documents/Reports/2021/2021%20Womens%20Initiative%20Declaration.pdf?ver=2021-06-10-194645-207

References

ABTA. (2022). *Travel in 2023 report.* [Online]. Retrieved February 4, 2022, from https://www.abta.com/sites/default/files/media/document/uploads/ABTA%20Travel%20in%202023%20report.pdf
Bae, S. Y., & Chang, P. J. (2020). The effect of coronavirus disease-19 (COVID-19) risk perception on behavioral intention towards "untact" tourism in South Korea during the first wave of the pandemic. *Current Issues in Tourism.* https://doi.org/10.1080/13683500.2020.1798895
Becken, S. (2004). How tourists and tourism experts perceive climate change and carbon offsetting schemes. *Journal of Sustainable Tourism, 12*(4), 332–345.
Becken, S., Simmons, D. G., & Frampton, C. (2003). Energy use associated with different travel choices. *Tourism Management, 15*(4), 257–278.
Beirman, D. (2022). *Tourism crises & destination recovery.* Sage.
Bhutan Department of Tourism. (2021). *Tourism policy.* [Online]. Retrieved February 23, 2023, from https://www.tourism.gov.bt/about-us/tourism-policy
Bhutan Travel. (2023). *Bhutan's sustainable development Fund.* [Online]. Retrieved July 18, 2023, from https://bhutan.travel/journal/editorial/bhutan-s-sustainable-development-fee

Bhutia, P. D. (2023a). *Bhutan offers duty-free gold to offset tourism fee impact.* [Online]. Retrieved July 22, 2023, from https://skift.com/2023/03/22/bhutan-offers-duty-free-gold-to-offset-tourism-fee-impact/

Bhutia, P. D. (2023b). *Bhutan lowers sustainable fee to boost longer stays from tourism.* [Online]. Retrieved July 18, 2023, from https://skift.com/2023/06/22/bhutan-lowers-sustainable-fee-to-boost-longer-stays-from-tourists/

Bordeaux Tourism and Conventions. (2023). *Act for a sustainable tourism.* [Online]. Retrieved January 20, 2023, from https://www.bordeaux-tourism.co.uk/act-sustainable-tourism

Borges, A. P., Vieira, E., & Gomez, S. (2020). The evaluation of municipal tourist tax awareness: The case of the City of Porto. *Tourism and Hospitality Management, 26*(2), 381–398. https://doi.org/10.20867/thm.26.2.6

Cetin, G., Alrawadieh, Z., Dincer, M. Z., Dincer, F. I., & Ioannides, D. (2017). Willingness to pay for tourist tax in destinations: Empirical evidence from Istanbul. *Tourism Economics, 5*(2), 21. https://doi.org/10.3390/economies5020021

Dodds, R., & Butler, R. W. (Eds.). (2019). *Overtourism: Issues, realities and solutions.* Walter De Gruyter.

Dolnicar, S., & Grün, B. (2009). Environmentally friendly behavior: Can heterogeneity among individuals and contexts/environments be harvested for improved sustainable management? *Environment and Behavior, 41*(5), 693–714.

Dolnicar, S., & Zare, S. (2020). COVID-19 and Airbnb—Disrupting the disruptor. *Annals of Tourism Research, 83*, 102961. https://doi.org/10.1016/j.annals.2020.102961

Edgell, D. (2016). *Managing sustainable tourism: A legacy for the future.* The Haworth Hospitality Press.

Edgell, D. L., & Swanson, J. R. (2019). *Tourism policy and planning: Yesterday, today and tomorrow* (3rd ed.). Routledge.

Goodwin, H. (2011). *Taking responsibility for tourism.* Goodfellow Publishers.

Gunderson, L. H., Allen, C. R., & Holling, C. S. (Eds.). (2012). *Foundations of ecological resilience.* Island Press.

Hall, C. M. (2008). *Tourism planning policies, processes and relationships* (2nd ed.). Pearson.

Hall, C. M. (2018). Resilience theory and tourism. In *Resilient destinations and tourism* (pp. 34–47). Routledge.

Hjalager, A.M. (2018). Suppliers as key collaborators for sustainable tourism. In J. Liburd & D. Edwards (Eds.), *Collaboration for sustainable tourism development.* Goodfellow Publishers.

Holloway, J. C., & Humphreys, C. (2022). *The business of tourism* (12th ed.). Sage.

Instat (Institute of Statistics Albania). (2023). *Arrivals of foreigners in Albania by purpose of travel, 2014–2022*. [Online]. Retrieved July 22, 2023, from https://www.instat.gov.al/en/themes/industry-trade-and-services/tourism/#tab2

International Labour Organization. (2022). *The gender gap in employment: What's holding women back?* [Online]. Retrieved July 18, 2023, from https://www.ilo.org/infostories/en-GB/Stories/Employment/barriers-women#intro

International Monetary Fund (IMF). (2020). *Wish you were here*. [Online]. Retrieved February 24, 2023, from https://www.imf.org/en/Publications/fandd/issues/2020/12/impact-of-the-pandemic-on-tourism-behsudi

International Monetary Fund (IMF). (2021). *Tourism in a post pandemic world*. [Online]. Retrieved February 24, 2023, from https://www.imf.org/en/News/Articles/2021/02/24/na022521-how-to-save-travel-and-tourism-in-a-post-pandemic-world

Lade, C., Strickland, P., Frew, E., Willard, P., Nagpal, S., Osorio, S. C., & Vitartas, P. (2020). *International tourism futures: The drivers and impacts of change*. Goodfellow Publishers.

Liburd, J., & Edwards, D. (2018). *Collaboration for sustainable tourism development*. Goodfellow Publishers.

Mashpi Lodge. (2023). *About the lodge*. [Online]. Retrieved February 23, 2023, from https://www.mashpilodge.com/the-lodge/?utm_medium=ppc&utm_source=Google&utm_campaign=ml_searchnontm_campana_continua&utm_sp=B2Cint&gclid=EAIaIQobChMIhsLO8oms_QIVgfftCh3uOg8oEAAYASABEgIbD_D_BwE

Melissen, F., & Sauer, L. (2019). *Improving sustainability in the hospitality industry*. Routledge.

Metz, B., Davidson, O. R., Bosch, P. R., Dave, R., & Meyer, L. A. (Eds.). (2007). *Mitigation of climate change—Contribution of working group III to the fourth assessment report of the intergovernmental panel on climate change*. Cambridge University Press.

Milano, C., Cheer, J. M., & Novelli, M. (Eds.). (2019). *Overtourism: Excesses, discontents and measures in travel and tourism*. CABI.

National Geographic. (2021). *Can science solve the seaweed problem on Mexican beaches?* [Online]. Retrieved February 23, 2023, from https://www.nationalgeographic.com/travel/article/can-science-solve-the-seaweed-problem-on-mexican-beaches

OECD. (2020). *OECD tourism trends and policies 2020.* OECD Publishing. https://doi.org/10.1787/6b47b985-en
Pinke-Sziva, I., Smith, M., Olt, G., & Berezvali, Z. (2019). Overtourism and the night-time economy: A case study of Budapest. *International Journal of Tourism Cities, 5*(1), 1–16. https://doi.org/10.1108/IJTC-04-2018-0028
Republic of Albania Ministry of Tourism and Environment. (2018). *National strategy for sustainable tourism development 2019–2023.* [Online]. Retrieved July 22, 2023, from https://turizmi.gov.al/wp-content/uploads/2019/12/National-Tourism-Strategy-2019-2023-EN.pdf
Suresh, K., Wilson, C., Quayle, A., Managi, S., & Uttam, K. (2022). Can a tourist levy protect National Park Resources and compensate for wildlife crop damage? An empirical investigation. *Environmental Development, 42*(2022). https://doi.org/10.1016/j.envdev.2021.100697
Tourism Authority of Thailand. (2022a). *Press release: TAT strengthens "Visit Thailand Year 2023" towards a meaningful travel direction.* [Online]. Retrieved January 19, 2023, from https://www.tatnews.org/2023/01/tat-strengthens-visit-thailand-year-2023-towards-a-meaningful-travel-direction/
Tourism Authority of Thailand. (2022b). *Press release: TAT debuts 'meaningful travel: Responsible and sustainable tourism' at WTM 2022.* [Online]. Retrieved January 19, 2023, from https://www.tatnews.org/2022/11/tat-debuts-meaningful-travel-responsible-and-sustainable-tourism-at-wtm-2022/
Üngüren, E., & Güçlü, C. (2021). COVID-19 pandemic and the hospitality and tourism industry: Effects on socio-cultural environment. In D. Gursoy, M. Sariisik, R. Nunkoo, & E. Bogan (Eds.), *Covid-19 and the hospitality and tourism industry: A research companion.* Edward Elgar Publishing Limited.
United Nations. (2022). *The sustainable development goals report 2022.* [Online]. Retrieved February 4, 2023, from https://unstats.un.org/sdgs/report/2022/The-Sustainable-Development-Goals-Report-2022.pdf
United Nations. (2023a). *The 17 goals.* [Online]. Retrieved February 26, 2023, from https://sdgs.un.org/goals
United Nations. (2023b). *Speech by UN women executive director Sima Bahous.* [Online]. Retrieved February 26, 2023, from https://www.linkedin.com/feed/update/urn:li:activity:7034099473282658304/
UNWTO. (2018). *Tourism data dashboard: Tourism's contribution to Thailand's GDP.* [Online]. Retrieved February 24, 2023, from https://www.unwto.org/tourism-data/global-and-regional-tourism-performance
UNWTO. (2022a). *Tourism for the SDGs.* [Online]. Retrieved January 19, 2023, from https://tourism4sdgs.org

UNWTO. (2022b). *SDG 5 gender equality.* [Online]. Retrieved January 19, 2023, from https://tourism4sdgs.org/sdg-5-gender-equality/

UNWTO. (2023a). *UNWTO tourism and the SDGs.* [Online]. Retrieved July 18, 2023, from https://www.e-unwto.org/doi/book/10.18111/9789284419401

UNWTO. (2023b). *UNWTO tourism recovery tracker.* [Online]. Retrieved July 22, 2023, from https://www.unwto.org/tourism-data/unwto-tourism-recovery-tracker

Visit Greenland (2021). *Towards More Tourism.* Accessed via https://travel-trade.visitgreenland.com/wp-content/uploads/2021/02/Strategi-EN-feb2021.pdf

West Sweden Tourist Board. (2023). *About stepping up sustainability.* [Online]. Retrieved July 22, 2023, from https://hallbarhetsklivet.se/en/

Wiki Commons. (1991). *Expedition trails surrounding Mashpi Lodge.* https://commons.wikimedia.org/

Wiki Commons. (2022). *Gender equality and empower women and girls.* https://commons.wikimedia.org/

World Economic Forum. (2022). *The promise and challenges of sustainable travel products.* [Online]. Retrieved February 24, 2023, from https://www.weforum.org/agenda/2022/11/sustainable-travel-and-tourism/

WTTC. (2018). *Tourism provides jobs where they're needed most.* [Online]. Retrieved February 23, 2023, from https://worldtraveltourismcouncil.medium.com/tourism-provides-jobs-where-theyre-needed-most-200f11ac2029

WTTC. (2021). The Cancun women's initiative declaration supporting women in travel & tourism: Global declaration and platform: Women's equality and leadership. [Online]. Retrieved February 9, 2023, from https://wttc.org/Portals/0/Documents/Reports/2021/2021%20Womens%20Initiative%20Declaration.pdf?ver=2021-06-10-194645-207

WTTC. (2022). *Travel and tourism economic impact 2022.* [Online]. Retrieved February 23, 2023, from https://wttc.org/Portals/0/Documents/Reports/2022/EIR2022-Global%20Trends.pdf

12

Exploring Underwater Tourism in EU-Mediterranean Coastal Areas for Sustainable Development

Sofia Gkarane and Chris A. Vassiliadis

Introduction

Many of the EU-Mediterranean coastal areas are an amalgam of cultural and natural extraordinary resources. Thus, there is a growing trend towards the economic and social impact of developing coastal, maritime, and underwater tourism in Europe. The significance of this trend was underscored during the European Maritime Day in 2022, which spotlighted its potential to foster a sustainable blue economy for green recovery. Despite its promise, empirical studies on the role of underwater tourism opportunities and their contribution to sustainable challenges remain scarce. In response to this research gap, the present study examined the impact of underwater tourism development to the coastal areas, through a qualitative method based on worldwide best practices. Data were collected using desk research and secondary databases. This study contributes to the existing knowledge of multidisciplinary research

S. Gkarane (✉) • C. A. Vassiliadis
University of Macedonia, Thessaloniki, Greece
e-mail: sgarane@uom.edu.gr; chris@uom.edu.gr

N. Stylos et al. (eds.), *Contemporary Marketing Management for Tourism and Hospitality*, https://doi.org/10.1007/978-3-031-65049-9_12

interests such as new perspectives for coastal area tourism destination development that could ensure a conservation management of the underwater historical and natural attractions from the part of the local authorities, for the education and benefit of the local communities. The aim is to raise awareness of new innovative practices and technology of life under water both by confirming the beneficial contribution of such practices and by providing valuable managerial insights for practical implications. Finally, the analysis can serve as a valuable tool for useful managerial decisions about underwater related tourism practices, aligned with the Sustainable Development Goal number 14—Life Below Water, as part of the United Nations (UN) 2030 Agenda.

The Mediterranean seabeds hold a captivating allure for water enthusiasts, offering them the chance to dive and explore the captivating underwater world using specialized equipment like oxygen cylinders. Underwater tourism is a special form of tourism with significant scope for development for the Mediterranean countries. Notably, Greece unveiled its first underwater museum in 2020, situated off Alonissos on the coast of the island of Peristera. This archeological site, accessible to recreational divers under expert guidance, holds immense historical significance as one of the oldest shipwrecks in the world, making it comparable in uniqueness to the renowned Parthenon. The Peristera wreck has placed Greece on the global diving tourism map, establishing the country as a prominent destination for underwater exploration.

Beyond Alonissos, the Greek seas harbor a wealth of submerged treasures, alluring amateur divers with over 1500 shipwrecks from various eras and sunken cities dating back to prehistoric times. Regions like Messinia and Pagasitikos boast numerous diving centers, while Crete, Corfu, and Piraeus offer an abundance of diving and training facilities, surrounded by pristine Mediterranean waters.

The establishment and exploitation of "underwater dive parks," similar to the one mentioned, present a compelling opportunity for fostering innovative dive tourism activities. Several Greek regions house deep archeological discoveries, primarily at shallow depths, serving as the foundation for implementing this concept. Thus, the main focus of this chapter is to explore the impact of underwater tourism development on Mediterranean coastal areas. Additionally, this chapter delves into fresh perspectives on coastal area tourism destination development, ultimately affirming the positive contribution of innovative underwater practices and technology to enriching the experience of life under water.

Background

It has been close to 150 years since the publication of the work of French writer Jules Verne entitled, "Twenty Thousand Leagues Under the Seas: A Tour of the Underwater World" which prophesied his invention of submarine and exploration of the seabed; and about 85 years after the invention of the first scuba diving mask made by the French oceanographer Jacques-Yves Cousteau with the motive of exploring the oceans. This silent world (the seabed) has prompted for decades, the interest of readers who saw through the pages of the Jacques-Yves Cousteau Books, an inexhaustible natural environment that hides within it, rich natural and cultural materials for processing and photography. Jacques-Yves Cousteau, the founder of the Advanced Marine Research Center, with the ex-naval and exploratory craft Calypso, began in 1955 to record with his camera and archive for the later researchers, this unique undersea wealth of our planet. Governments of the post-war era did not realize this new perspective of natural and cultural exploration as well as the possibility of exploiting and promoting this heritage. The example of the Greek government that in 1975 invited Jacques-Yves Cousteau to explore the area of Santorini in the Aegean and the possible discovery of the Lost Atlantis is typical (ERT, 2023).

Nowadays, the issue of the use of diving destinations is becoming more and more relevant. Exploiting the potential of the relevant coastal destinations and making them eligible as distinct destinations with unique characteristics that differentiate them from other competing destinations are important points of reference in relevant studies (Acorn Consulting Partnership Ltd, 2008, 2014). The operators of the geographical destinations appropriately shaped the policies of attracting investors, hosting and developing diving and underwater tourism infrastructure, and highlighting these opportunities in order to attract stakeholders (investors and visitors). The attraction of special groups of visitors as well as investors who are interested in taking part in underwater tourism (such as diving underwater to meet the specific needs and preferences of those involved such as athletic, naturalistic, cultural tourism) is required due to the uniqueness and the dangers of the bio-changes in the underground natural environment in other to safeguard or accredit participants' competencies by specialists, professionals or institutional bodies. The licensing and validation of the capabilities of the participants (clients and organizers) are recognized internationally by certifications from international organizations such as Professional Association of Diving Instructors (PADI, 2018) founded in 1966; National Association of Underwater Instructors (NAUI, n.d.) established in 1959; Scuba Schools International (SSI, n.d.) established in 1970 in the United States; British Sub-Aqua Club (BSAC, n.d.) founded in 1953; Confederation Mondiale des Activities Subaquatiques (CMAS, n.d.) established in 1959; VDST (VDST, 2023; Association of German Scuba Divers) established in 1954; and many other diving schools within European headquarters. The demand and the related development of diving activities are growing internationally, with the result that investors are increasingly interested in exploiting this emerging market (Uyarra et al., 2009). According to PADI's certified diversified statistical analysis—which is also considered to be the market leader and provides 70% of worldwide certifications—by the end of 2012, the organization had allocated over 21.3 million certifications. Based on forecasts from the Mintel market analyst, the Market is dominated by about 40 million divers worldwide, of which certified divers amount to about 30 million people. This World Market includes about 10 million active divers (the Global Market is about 1 million active divers). The largest deer shipping markets are the US, Italy, Germany, France, and the UK. Participating divers prefer holiday travels of seven to ten days on the average and an average cost of $2500. The diving retail market creates jobs in coastal areas. An example is Noumeri Nasr who lives in Egypt (Hurghada, in Egypt). His knowledge of English and German brought him close to tourists from an early age. Today he is a dive master and a company manager, while his family members and young locals (over 500 divers) now seem to be following the example of choosing the diving profession he also chose to work (Mendiratta-UNWTO, 2013). The coastal regions and particularly the

Mediterranean ones, that is, those that are concerned with this paper, benefit in terms of the quality of life and job creation. This development is often framed by sustainable development programs that are aimed at preserving and harnessing natural and cultural wealth as a wealth of heritage. These programs involve the member communities of EU member states—such as the MED Community on Sustainable Tourism project (BleuTourMed—Plan Bleu, n.d.)—often supported and integrated into the aspirations of wider international organizations such as the World Tourism Organization (WTO), under the supervision of the United Nations. The protection and visibility of natural and cultural resources in the Mediterranean region is an objective of these 17 programs—collaborations between clusters of projects (17 clusters of projects) within the framework of the Interreg Medium Program 2014—2014 (MED Sustainable Tourism Community, n.d.-a). Life below Water has been set as the 14th among the 17 Sustainable Development Goals (SDGs) of the United Nations for a better world for people and the planet by 2030 (Sustainable development goals (SDG 14): UN Western Europe, 2020). The aim of the 14th goal is to conserve and sustainably use the oceans, seas and marine resources. In line with the UN's aspirations, any development effort should internationally be stepped up and is based on the satisfaction of five key pillar criteria concerning the goals set for the United Nations by the United Nations (UN) for Sustainable Tourism and Development (IY2017). The pillars for the rational and sustainable development of tourism are as follows:

1. Sustainable economic growth;
2. Social inclusion, employment, and poverty reduction;
3. Resource efficiency, environmental protection, and climate change;
4. Cultural values, diversity, and heritage; and
5. Mutual understanding, peace, and security.

Defining Underwater Tourism

> Dive tourism involves travel with the primary objective of exploring the underwater attractions of a vacation destination. This typically entails scuba diving, utilizing specialized underwater breathing equipment, such

> as an oxygen cylinder and breathing regulator. The activities encompass diving in both freshwater and marine environments, while snorkeling in any form is not considered part of dive tourism. (Acorn Consulting Partnership Ltd, 2014)

> Marine tourism refers to leisure activities that entail traveling outside of one's usual place of residence and center around the marine environment. The marine environment is defined as waters that are saline and influenced by tides. (Orams, 1998)

Activities related to scuba diving, snorkeling, windsurfing, fishing, watching sea mammals and seabirds, all beach activities, Maritime museums, sailing and motor boating etc. are also listed in this definition (Martínez Vázquez et al., 2021).

> Marine tourism comprises a diverse array of tourism, leisure, and recreational activities that occur within the coastal zone and the adjoining offshore coastal water. (Orams & Lück, 2014)

> The concept of coastal tourism embraces the full range of tourism, leisure, and recreationally oriented activities that take place in the coastal zone and the offshore coastal waters. These include coastal tourism development (accommodation, restaurants, food industry, and second homes), and the infrastructure supporting coastal development (e.g. retail businesses, marinas, and activity suppliers). Also included are tourism activities such as recreational boating, coast- and marine-based ecotourism, cruises, swimming, recreational fishing, snorkeling and diving. (Hall, 2001)

The terminology of Underwater Tourism is based on the synthesis of two words and a well-defined definition is not easy to find until today. So, with the scope to define the activities introduced to our following desk research study, it is attempted to be expressed in the following paragraph.

Underwater tourism includes an amalgam of activities and more specifically the following: Underwater constructions like underwater hotels and accommodations, underwater restaurants and FB (food & beverages) facilities, underwater heritage parks like underwater museums, e.g., Sunken treasures, shipwrecks and storytelling activities for these places, coral reefs exhibitions, submarine touring and diving activities (i.e., Scuba, Snorkeling).

Legal Framework for the Development of Diving Activities

Underwater archeology can enhance the whole underwater tourist attractions. In the ancient times, merchant vessels were traversing from ancient port to ancient port through specific sea routes, i.e., from Anatolia, Samos, Black Sea region to Rhodes, Cyprus, and Egypt. Today archeologists have discovered a great number of wrecks by using new scuba, informatics, and underwater marine technology (Romeo, 2015). The Legal Framework and its coordination in the national and international level seems to be a complicated situation until today. The EU and Nationwide restrictions are about underwater activities and the need for the protection of the archeological or natural sites. However, the non-compatibility between their Laws are some of the problems and constraints of developing a unique underwater EU policy for tourism leisure activities (Dimmock & Musa, 2015; Romeo, 2015).

Underwater Tourism

The human activities that are a component of underwater tourism activities are obviously expanding into other activities that can be combined with diving tourism and expanding the diving tourism market. He integration of innovative technologies and pioneering experiments is bringing to life the visionary concepts of Jules Verne and Jacques-Yves Cousteau, fostering this development. Visitors' engagements now span across distinct areas, grouped into three categories, each offering unique and captivating exhibits:

(a) Controlled areas outside the natural environment (e.g., water parks, water pools, museums with underwater attractions);

(b) Natural spaces and within the environment (e.g., natural underwater coral parks, submarine parks for exploration of wrecks and underwater life); and
(c) Controlled areas with significant anthropogenic interventions within the physical space (e.g., controlled and secured areas such as protected subterranean parks that are delineated within the natural environment, underground accommodation infrastructure and underwater infrastructures for the presentation of the natural habitats and the environment).

A record of the important trends of underwater tourism in the Mediterranean region is mentioned below.

Underwater Accommodations and Infrastructures: Trends and Situation in the Mediterranean Region

Despite the variety of natural and cultural sites in the Mediterranean, so far there is no remarkable and reputed underwater accommodation. It is typical that the Internet only shows a European underwater accommodation on a relevant Skyscanner Ltd web site and this is located in a lake in Northern Europe. In Sweden, a local artist named Mikael Genberg built Utter Inn in Lake Malaren. It is a small accommodation that features an underwater room with panoramic view due to its side windows (Orams, 1998). As it becomes clear, the Mediterranean European Basin does not have [when compared to the areas that host exceptional underwater infrastructure such as Atlantis (The Palm-Dubai), The Manta Resort (Pemba Island-Zanzibar), The Underwater Hotel Club Subsix (Niyama-Maldives) Underwater Ithaa Restaurant (Conrad, Rangali Island-Maldives), Jules' Undersea Lodge (Key Largo, Florida), Lime Spa (Huvafen Fushi-Maldives), Lovers Deep (Caribbean), etc.] the capacity, variety and services aimed at attracting lucrative visitors. The accommodation rooms within an environmental mindset are often the good examples that are models for future investments.

One UNWTO housing case, as a best practices case study of demonstrating reductions in energy consumption by up to 70%, is the neZEH; Nearly Zero Energy Hotels project (UNWTO, 2018).

This green and hospitality friendly project is the implementation of ten co-partners from seven countries. The overall neZEH project is a project that uses more than 1540 hotels that are members of the Hotel Energy Solutions (HES) initiative. The hotels are members of this hotel cluster project with the scope of improving their competitiveness through resource efficiency. In the EU, until today, the implementation of renovation projects is a reality. Between 2013 and 2016, 16 hotels of this cluster project have supported and completed their renovation projects with many benefits for their operation cost and their sustainability, like 70% lower energy consumption than fewer. The ten partners mentioned above are as follows:

1. Technical University of Crete's Renewable and Sustainable Energy Systems Laboratory;
   which acts as neZEH's Coordinator;
2. World Tourism Organization (UNWTO);
3. Network of European Regions for Competitive and Sustainable Tourism (NECSTouR);
4. Federation of European Heating, Ventilation and Air Conditioning Associations (REHVA);
5. Agency of Brasov for the Management of Energy and the Environment (ABMEE Romania);
6. Creara Spain;
7. ENERGIES2050 France;
8. Energy Institute Hrvoje Požar (EIHP Croatia);
9. Italian Industrial Transmission Society (SiTI Italy); and
10. Sustainable Innovation (SUST Sweden).

Also a lot of supporters are involved in this cluster project, like, tourism stakeholders, including hotel associations, public authorities, policy makers and financial institutions at the national and EU levels. Hotel guests are also engaged to support energy saving. neZEH has contributed to support for energy efficiency and climate change policies that are related to 2020 and 2050 targets.

Underwater and coastal community lodging capacities can also be one of the final user(s) of these innovative hospitality beneficial actions of the neZEH consortium (Intelligent Energy Europe Programme; Technical University of Crete, 2013).

Archeological Tourism

The cultural routes that connect underground exhibits and archeological–historical findings among them are of particular interest in terms of exploitation potential. Computing and multimedia applications, combined with developments in telecommunications and satellite imaging systems, can make a significant contribution to the promotion of underwater archeological sites such as the first Human-Robot Interface with the title "Dolphyn" (Dominques et al., 2012) and project VISAS, which aimed at (among other things) the three-dimensional representation of underwater archeological sites in the coastal areas of Southern Italy. The area of the Mediterranean includes over 1000 submarine shipwrecks discovered by divers and involving the Classical and the Middle Ages.

Excursions and Exploration with Underwater Submarines

Next, some best innovative practices for underwater tourism are presented with the use of submarine technologies (Viator, 2023). The first case is the Submarine Adventure Seabed of Puerto de Mogan 40 minutes tour in the water area of Mogan Spain. The non-diving visitor discovers all the attractive elements of the seabed like shipwrecks and fauna. One of the underwater submarine practices that are related with the Med-area is also the case of the Yellow Submarine Mogan at Maspalomas of Spain. I.K. from Ireland wrote a review on trip advisor's web site with the following words:

> Really great experience, you get to see all sorts of aquatic life and shipwrecks around Puerto de Mogan. Would definitely recommend, especially with kids. (Trip Advisor; Customer review at 6 September 2018 via mobile)

The second case is the submarine dive excursion from Antalya and Kemer or Belek, Turkey; The 1 h tour for about 46 passengers with the submarine NEMO can create a fantastic underwater experience for the non-diving visitors. The Spanish submarine was renovated in 2015 and relocated to Antalya. G.B. is a visitor of the area and a passenger of this submarine; he or she expressed the following experience on trip advisor's web site:

> What an amazing experience. You go out about 1 km and then dive on a shipwreck. There are lots of small fish. They said the water was not very clear. But it was still worth the trip. A video explains in Turkish English and maybe German. It's great for all the family and older people. As long as you can climb a ladder. I would love to go again one day. (Trip Advisor; Customer review at 16 September 2017)

Turkey also has the advantage of organizing many underwater journeys in the eastern side of the Mediterranean Sea. Another good underwater tourism practice is the semi submarine tour in Marmaris area of Turkey. A UK visitor of the place and user of this submarine service described his experience with the following words:

> Just back from the day trip and we loved it, might go back before we go home. The underwater viewing glass is amazing to see the shipwrecks and the fish from below. Dinner was good and there was plenty time at all the stops, which were perfectly chosen. If looking for a boat tour I highly recommend this one. Full day price includes dinner around £16... Bargain! (Trip Advisor; Customer review at 6 July 2018)

Near the Montenegro Adriatic Coastline area; at Kotor Port area, families and day tourists can have a great experience using the semi submarine excursions service. G.S. from UK expressed his experience from his underwater trip with the following words:

> Family day out to Kotor we wanted to go on the water for an hour or so to have a leisurely look round the bay. Decided on the Red submarine, which was in the harbor by the park. We had a 1 year old baby and two boys with us and they said it was no problem, they weren't busy at the time so we negotiated a price for us all which they accepted. So we had the sub to ourselves. Nice dry weather both the boys got a go at steering the sub which they loved, it was great fun. Didn't see much below the water when we went down below for 15 minutes to look but the whole experience for the family was great. (Trip Advisor; Customer review at 21 April 2017)

Another good practice that is related to local authorities and underwater tourism development is the case of MuMART Museum underwater visit in a submarine at Golfo Aranci of Italy. This is a live water museum with modern and creative art in the bottom of the sea. The owner of the art attractions and of the semi-submersible submarine is the municipality of Golfo Aranci. The submarine dives between four and seven meters deep on the white seabed off the third beach of Golfo Aranci (Review of MuMart, Trip Advisor, n.d.).

Good Practices in Diving Tourism

- Alexandria Eastern Harbor Underwater Museum. The location chosen for the underwater museum is the Eastern harbor, with Greek, Roman, and Pharao attractions. The architectural design of the underwater museum got approved by the UNESCO and the Egyptian Ministry of Antiquities. The underwater museum is a live museum of Egypt's underwater treasures (Amira Elhamy, 2016).
- Italy (Baia). The Baia underwater area was discovered in 1959 and it's an underwater archeological park with five main diving areas. The most remarkable areas are the Ninfeo di Claudio and the Villa a Pprotiro. The divers can explore relation rooms of the Emperor, a paved road, statues like the original ones, mosaics, walls of buildings etc. (World Adventure Divers, n.d.).
- Diving for shipwreck lovers, underwater sea area of Malta; Island Gozo and the Blue Hole dive site. The area attracts both divers and non-divers

and family-oriented tourism (Padi Associates Environmental Consultants Ltd, 2010).

- Scuba diving sea park underwater sea area of Mendes, Spain. Benefits: Over 20,000 dive visitors visit the park annually; diving tourism brings non-seasonal and all-year-round tourism opportunities; scuba diving brings to the local economy three million Euros per year (Kipreos et al., 2016).

Museums in the Mediterranean Region

- Greece (Alonissos Underwater Museum). It is the first of its kind in Greece that is open to the public. The shipwreck is now open to recreational divers who will have the opportunity to explore it under the guidance of experienced divers.
- Egypt (The Project of Alexandria Underwater Museum). It is a live museum as presented before, in "Good practices in diving tourism" section.
- Spain (National Museum of Underwater Archeology ARQVA, Cartagena, Region of Murcia). A maritime attraction like a Roman merchant ship. The visitors can see many Phoenician, Galley, and Roman archeological collections like ancient ships from underwater archeological excavations. The emphasis is given to trading, maritime and transportation activities in different time periods (National Museum of Underwater Archaeology, n.d.).
- Turkey (Bodrum Museum) and organized by the Turkish Underwater Antiquity Officials diving opportunities to explore the underwater archeological attractions of the Bodrum area (Kipreos et al., 2016).

Pillars of Responsible and Sustainable Development

The study extensively examines the practices of underwater tourism (UWT) using the five pillars prism of the International Year of Sustainable Tourism for development (IY2017) (MED Sustainable Tourism

Community, n.d.-b) to contribute to the in-depth analysis of the UWT trend and practices.

In the following text, critical points for the evaluation of the above trends and practices in underwater tourism activities are presented.

Pillar One: Sustainable Economic Growth

Coastal and underwater tourism are important for the economic (business) and social (e.g., quality of life, young population) development of coastal areas. The sea-based attractions depend on the status and the form of human activities in the area; and the need for preservation of the maritime environment can also bring a stable economic perspective to this area (sustainable economic growth). The business of marine recreational activities such as swimming, diving and fishing depend on the water quality status; and a negative water quality status (i.e., water pollution) can also affect the income and business perspectives of the local economy (Tore Söderqvist et al., 2012).

Pillar Two: Social Inclusiveness, Employment, and Poverty Reduction

Total employment in the EU Med professional fisheries sector has been estimated at around 250,000 jobs in 2008 (Sacchi, 2011). The majority of EU coastal areas are villages and small regional or local city centers with populations that have low-income status; and the need for job creation and non-seasonal job places is more important than in other areas (Orams, 1998).

Pillar Three: Resource Efficiency, Environmental Protection, and Climate Change

Because of the great variety of anthropogenic factors that can negatively affect the biodiversity and the status of the underwater resources as well as the related tourist attractions, it is important to enforce environmentally friendly actions and behaviors e.g., actions to avoid taking during diving and underwater activities, the possible damage of corals, reef damages and changes in fish variety and quantity in the underwater parks, and the underwater cultural heritage (Shackley, 1998; Tore Söderqvist et al., 2012; Uyarra et al., 2009). The new EU strategy for a Sustainable Blue Economy, has been a recent new approach for a transformation of its economy to become more sustainable for future generations and to encompass all industries and sectors related to oceans, seas and coasts. In the last decades, areas with marine ecosystems that are affected by the rising sea temperature can be defined in the Mediterranean Sea. Fishes and planktons are now proliferating (spreading rapidly) also in the Northern and Western Mediterranean Sea regions. These kinds of changes can seriously affect the EU population growth, tourism development and the attractiveness of coastal areas (Piante & Ody, 2014). The Mediterranean Sea does not have coral reefs like other parts of the world oceans. But the Mediterranean Sea still contains a different kind of underwater life. About 40% of marine species in the Mediterranean underwater areas are native to the region and unique worldwide (Mediterranean Snorkeling Report, n.d.).

Pillar Four: Cultural Values, Diversity, and Heritage

In relation to environmental conservation underwater strategy, the management of natural resources can usually bring the chance to protect local heritage or to revitalize native cultures. For instance, by regenerating traditional arts and crafts and their underwater and local cultural heritages like ancients' shipwrecks and old ports in the Mediterranean Sea (see also Fig. 12.1) (UNESCO: The Protection of the Underwater Cultural Heritage, 2001).

Pillar Five: Mutual Understanding, Peace, and Security

Intercommunication between different cultures of people is important for tourism development. Cultural events, underwater related story telling, diving and fishing festivals, small-scale sport events can contribute to

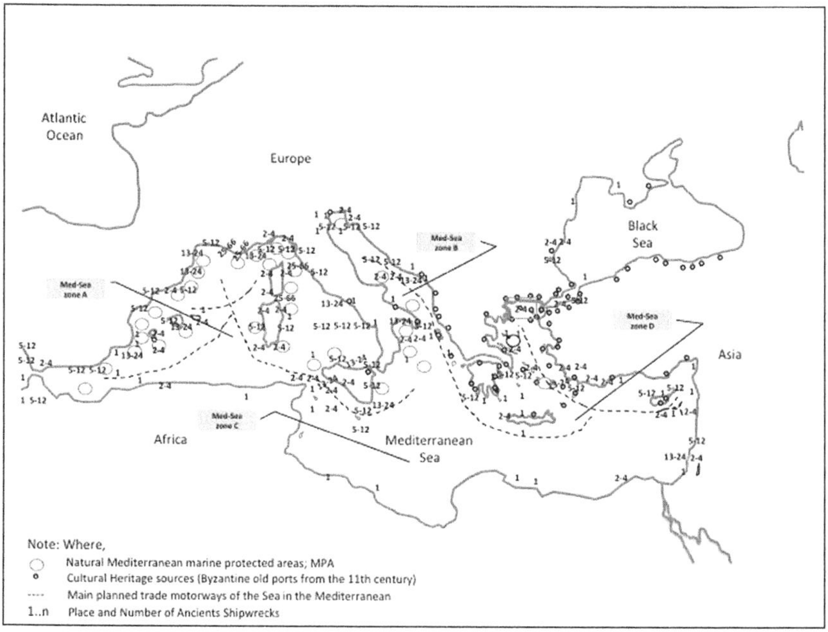

Fig. 12.1 Med diving opportunities and protected areas. Location of Mediterranean Protected Areas, the trend of the Sea Cargo Trade Flows and Archeological Ancient underwater attractions with Byzantine old ports of eleventh century. (Source: Authors)

the pillar actions of mutual understanding, peace and security (UNWTO, 2017a, b; 2018).

The pillars above are important for the identification of a unique strategy through effective mapping of all the five pillars e.g., with emphasis on, attractions, protected and non-protected areas, seaside human underwater and water leisure activities, impact, and water pollution activities of humans in the seaside, water and underwater areas, humanity, cultural effects and bringing civilizations together. In this study, a proposal map with natural and cultural Med-areas that can be a basis for more in depth research about diving and underwater tourism activities in the Med-area is presented in Fig. 12.1.

As can be identified from the spatial analysis of the Med-Sea area, there are zones that can be difficult to manage because of the internationally protected natural and cultural areas. The coastlines of the African

Med-geographical marine water spaces can be an easy policy scenario for the development of underwater tourism activities and the connection of the projects and policies between investors, project clustering members and administration bodies. The scenario can be more complicated for a unique managerial international project that is located on coastal waters and have overlapping spaces of protected natural and cultural water (see Fig. 12.1; Sea zones A, B, D). Such legal and regulation oriented (but policy and strategy management complicated) geographical waters can be a separate group of waters. Figure 12.1 presents a map of Med coastal and water areas with information about cultural and natural sources. The following Med water zones can be identified in Fig. 12.1:

- *Sea zone A* (i.e., (a) South east Spain, (b) South France, and (c) West Italy, coastal regions and water areas)
- *Sea zone B* (i.e., (a) Eastern Italy, (b) Southern-west Slovenia, (c) Western Croatia, (d) Southern-west Bosnia, (e) Southern Montenegro, (f) Western Albania and (g) Western Greece, coastal lines and related water areas)
- *Sea zone C* (i.e., (a) Northern Morocco, (b) Northern Algeria, (c) North-east Tunis, (d) Northern Libya, and (e) Northern Egypt, coastal regions and water areas)
- *Sea zone D* (i.e., (a) Mainly eastern areas of Greece, (b) Western areas of Turkey, (c) Cyprus, (d) West Syria, (e) West Lebanon, and (f) West Israel, coastal regions and water areas).

The identification procedure for the next zone includes (see Table 12.1) the economic analysis of growth opportunities in the Mediterranean area. Based on OECD: Regional Economy data (OECD, 2018), the situation of growth opportunities in these four Zone Areas can be identified.

More specifically, the Sea Zone A includes the following regions with their related growth indexes:

(a) Spain Med-Sea zone A: Catalonia (123.8), Valencian community (117.3), Murcia (130.8), Andalusia (122.5), Balearic (123.4) and Canary (115) Islands;

Table 12.1 Proposed future 5-pillar actions for each Mediterranean zone-area (A until D)

(**) Zone-Area: Pillar (*):	A	B	C	D
1. Sustainable Economic Growth	Ok; keep working on cluster(s) of projects based on EU regulations Use new trends and innovations	EU Development, unique regulations and more underwater job opportunities for the east sea-coastline Use new trends and innovations	Med-Synergies with projects and infrastructure for underwater and cultural tourism activities. Reinforce projects for cooperation Use new trends and innovations	Enforce Eastern- Med partnerships and the cooperation(s) for economic water tourism activities Use new trends and innovations
2. Social inclusiveness, employment and poverty reduction	Social inclusiveness, employment	Social inclusiveness, employment	Poverty reduction and employment	Social inclusiveness of Refuges, Poverty reduction and employment
3. Resource efficiency, environmental protection and climate change	Regulations and environmental protection	Resource efficiency, environmental protection	Resource efficiency, environmental protection	Regulations and environmental protection
4. Cultural values, diversity and heritage	Underwater and coastline cultural heritage protection. Plan a Med-EU member road map for co-cultural promotion	Underwater and coastline cultural heritage protection. Plan a Med-EU member road map for co-cultural promotion	International Regulations and agreements for Underwater and coastline cultural heritage protection	International Regulations and agreements for Underwater and coastline cultural heritage protection

5. Mutual understanding, peace and security	Keep peace and security in the area	Keep peace and security in the area	Very important Pillar for the Area (Stop war and Terrorism). Build relations between human populations and the different cultures	Very important Pillar for the Area (Stop war and Terrorism). Build relations between human populations and the different cultures

Note: (*) Where, 1–5=The five Pillars. (**) Where, A=Sea zone A (i.e., (a) South east Spain, (b) South France, and (c) West Italy, coastal regions and water areas), B=Sea zone B (i.e., (a) Eastern Italy, (b) Southern-west Slovenia, (c) Western Croatia, (d) Southern-west Bosnia & Herzegovina, (e) Southern Montenegro, (f) Western Albania and (g) Western Greece, coastal lines and related water areas), C=Sea zone C (i.e., (a) Northern Morocco, (b) Northern Algeria, (c) North-east Tunis, (d) Northern Libya, and (e) Northern Egypt, coastal regions and water areas), & D=Sea zone D (i.e., (a) Mainly eastern areas of Greece, (b) Western areas of Turkey, (c) Cyprus, (d) West Syria, (e) West Lebanon, and (f) West Israel, coastal regions and water areas)

(b) France Med-Sea Zone A: Occitanie (124.8), Provence-Alpes-Cote d' Azur (102.8), Corsica (118.9); and
(c) Italy Med-Sea Zone A: Liguria (94.2), Toscana (101.7), Lazia (99.1), Campania (94.5), Calabria (91.4), Sicilia (90.6), Sardinia (96).

In the same manner, the Sea Zone B:

(a) Italy Med-Sea zone B: Friuli-Venezia Giulia (96.1), Veneto (102), Emilia-Romagna (106.1), Marche (97.7), Abruzzo (97.3), Molise (85.2), Puglia; Apulia (92.5);
(b) Slovenia Med-Sea zone B: Obalno-Kraska, Coastal-Karst (101.1);
(c) Croatia Med-Sea zone B: Adriatic Croatia (131.3);
(d) Bosnia and Herzegovina (103.2);
(e) Montenegro: 4.5% (annual Real GDP growth; https://www.ceicdata.com/en/indicator/montenegro/real-gdp-growth);
(f) Albanian Med-Sea zone B: Shkodra (0.18) (2015 real growth rate contribution), Lezha (0.03), Durres (0.76), Tirana (2.10), Fier (-0.81), Vlora (-0.05) [http://www.instat.gov.al/media/1793/c-users-insalsp-desktop-webi-i-ri-national-account-angl-publikime-gross-domestic-product-by-statistical-regions-nuts-in-albania-year-2015.pdf]; and
(g) Greece Med-Sea zone B: Epirus (75.2), Western Greece (68.6), Peloponnese (77.2), Ionian Islands (70.3).

The third Sea Zone C includes the following countries or regions:

(a) Morocco (99.16, index 2005=100; https://www.economy.com/morocco/indicators);
(b) Algeria 1.30% per year (GDP Annual Growth Rate, 2018; https://tradingeconomics.com/algeria/gdp-growth-annual);
(c) Tunis 1.0% (per year GDP Annual Growth Rate, 2018; https://tradingeconomics.com/);
(d) Libya 55% (per year GDP Annual Growth Rate, 2018; https://tradingeconomics.com/); and
(e) Egypt 4.1% (yearly real growth rate in % of GDP 2017; https://www.indexmundi.com/g/g.aspx?c=eg&v=66).

Finally, the Sea Zone D includes the following countries or regions:

(a) East Macedonia and Thrace (74.1), Central Macedonia (73.4), Thessaly (75.4), Central Greece (73.8), Attica (73.3), Crete (76), South Aegean (75.4), North Aegean (71.8);
(b) Turkey (7.4% https://www.ceicdata.com/en/indicator/turkey/real-gdp-growth);
(c) Cyprus (98.4);
(d) Syria (-9.9%) (Yearly real growth rate in % of GDP 2015; https://www.indexmundi.com/g/g.aspx?c=sy&v=66);
(e) Lebanon (1.5%) (Yearly real growth rate in % of GDP 2017; https://www.indexmundi.com/g/g.aspx?c=sy&v=66); and
(f) Israel (3.1%) (Yearly real growth rate in % of GDP 2017; https://www.indexmundi.com/g/g.aspx?c=sy&v=66).

Conclusion

The EU policy on underwater tourism development seems to be a reality in these tech and e-turbulent days and EU has recognized the importance of underwater tourism in sustainable development in the EU Blue Growth Agenda (Dimitrovski et al., 2021). Environmental assessment and the need for more integrated underwater zone management strategies and practices are ongoing areas for discussion and sustainable tourism development and management, in the Med Sea Area. Private sector, authorities, EU – country, regional and local authorities (but also the local community with their representatives and international cluster of projects) must also be the basis for future actions in this goal. This desk research identifies the main areas that can contribute to this spatial integration of the Med Coastal Areas. Aspects that are related to the development(s) in tourism, underwater innovations, best practices and new EU strategies and natural and cultural underwater conservation perspectives have been presented and discussed (Bruno et al., 2017; Hall, 2018) (see Fig. 12.1). The main points of this desk research contribution are, to explain and to identify the aim of the description of European

Landscape and Underwater Tourism opportunities that are related with Med-tourism destination area(s) (Parker, 1987). The chapter highlights the importance of conducting more comprehensive research and evaluation to understand the impact of underwater tourism activities on coastal and underwater areas, i.e., identification of strict natural reserve underwater and marine areas, wilderness areas; national park areas; natural and cultural monuments; protected seascape and coastal landscape areas; human made attraction areas (Strati, 2006). Based on the above critical aspects for sustainable underwater tourism development, the spatial and economic analyses of four basic zones are crucial decision tool for future actions that can contribute to this goal (see also Table 12.1). A critical success point of the Med-strategy and policy on marine areas is to continue and to enforce the EU partnership actions, projects and activities in the future. Each area has in the Med-area, a different economic growth dynamic for Mediterranean underwater strategy and the related goal development policies. Also, the Med-Sea area political and legal integration policy need to be a first priority of all the Med-related countries (members and non-members of EU), with the scope to enforce (in a long-term manner) the Med-Strategy and policy for sustainable underwater tourism growth opportunities.

Recommended Further Reading

Cavallini, I., Marzo, D., Scaccia, L., Scipioni, S., & Niccolini, F. (2023). Scuba diving tourism and the challenge of sustainability: Evidence from an explorative study in North African-Mediterranean countries. *EuroMed Journal of Business.*

Loctier, D. (2023, July 28). *Virtual reality and ecotourism: Exploring and preserving our underwater heritage; From 3D models to virtual reality headsets and android robots, Ocean takes a deep dive into European projects using technology to forge a stronger connection between the public and our underwater cultural heritage.* European Commission, EuronewsGreen. https://www.euronews.com/green/2023/07/25/photogrammetry-virtual-reality-and-ecotourism-exploring-and-preserving-our-underwater-heri

Lucrezi, S., Ferretti, E., Milanese, M., Sarà, A., & Palma, M. (2021). Securing sustainable tourism in marine protected areas: Lessons from an assessment of scuba divers' underwater behavior in non-tropical environments. *Journal of Ecotourism, 20*(2), 165–188.

Manglis, A., Fourkiotou, A., & Papadopoulou, D. (2020). The Accessible Underwater Cultural Heritage Sites (AUCHS) as a sustainable tourism development opportunity in the Mediterranean Region. *Tourism: An International Interdisciplinary Journal, 68*(4), 499–503.

Papageorgiou, M. (2016). Coastal and marine tourism: A challenging factor in Marine Spatial Planning. *Ocean & Coastal Management, 129*, 44–48.

Sumarmi, S., Ensiyawatin, A. Z., Astina, I. K., Kurniawati, E., & Shrestha, R. P. (2022). The management of Bangsring Underwater as conservation-based ecotourism for international tourism destination. *GeoJournal of Tourism and Geosites, 41*(2), 393–399.

References

Acorn Consulting Partnership Ltd. (2008, February). *Developing a niche tourism market database for the Caribbean; 20 niche market profiles.* Caribbean Tourism Organisation. Retrieved November 28, 2022, from https://www.onecaribbean.org/wp-content/uploads/NicheMarketsDatabase.pdf

Acorn Consulting Partnership Ltd. (2014, March). *Dive tourism-South Pacific niche market report.* South Pacific Tourism Organisation. Retrieved November 28, 2022, from http://samoatourism.org/Content/SiteResources/PAGE/135/Dive%20Report.pdf

Amira Elhamy, A. (2016, February 2). *Underwater museum: Rebuilding the glory of sunken antiquities in Alexandria.* Meobserver.org. Retrieved November 28, 2022, from http://www.meobserver.org/?p=3762

Anita Mendiratta-UNWTO. (2013). *Tourism stories-how tourism enriched my life, Noumeri Nasr – Diving into the future.* UNWTO 2013 (pp. 57–59). Retrieved November 28, 2022, from https://www.e-unwto.org/doi/pdf/10.18111/9789284415571

BleuTourMed – Maritime and Coastal Sustainable Tourism in the Mediterranean, Plan Bleu. (n.d.). Building the Mediterranean future together. Retrieved November 28, 2022, from http://planbleu.org/en/activites/tourisme/bleutourmed-maritime-and-coastal-sustainable-tourism-mediterranean

Bruno, F., Lagudi, A., Barbieri, L., Muzzupappa, M., Mangeruga, M., Pupo, F., Cozza, M., Cozza, A., Ritacco, G., Peluso, R., & Tusa, S. (2017). *Virtual diving in the underwater archaeological site of Cala Minnola*. The International Archives of the Photogrammetry, Remote Sensing and Spatial Information Sciences, Volume XLII-2/W3, 2017 3D Virtual Reconstruction and Visualization of Complex Architectures, March 1–3, Nafplio, Greece. Retrieved November 28, 2022, from https://www.int-arch-photogramm-remote-sens-spatial-inf-sci.net/XLII-2-W3/121/2017/isprs-archives-XLII-2-W3-121-2017.pdf

BSAC, British Sub-Aqua Club. (n.d.). Retrieved November 28, 2022, from https://www.bsac.com/home/

CMAS. (n.d.). Confederation Mondiale des Activities Subaquatiques. Retrieved November 28, 2022, from http://www.cmas.org

Dimitrovski, D., Lemmetyinen, A., Nieminen, L., & Pohjola, T. (2021). Understanding coastal and marine tourism sustainability – A multi-stakeholder analysis. *Journal of Destination Marketing & Management, 19*, 100554.

Dimmock, K., & Musa, G. (2015). Scuba diving tourism system: A framework for collaborative management and sustainability. *Marine Policy, 54*, 52–58. https://doi.org/10.1016/j.marpol.2014.12.008

Dive in history: Underwater Pompeii in Pozzuoli, Italy; The Underwater Archaeological Park of Baia: An underwater Pompeii. (n.d.). World Adventure Divers. Retrieved November 28, 2022, from https://worldadventuredivers.com/underwater-pompeii-pozzuoli-italy/

Dominques, C., Essabbah, M., Cheaib, N., Otmane, S., & Dinis, A. (2012). *Human-robot-interfaces based on mixed reality for underwater robot teleoperation*. 9th IFAC Conference on Manoeuvring and Control of Marine Craft, 2012. The International Federation of Automatic Control, September 19–21. Arenzano, Italy. Retrieved November 28, 2022, from https://ac.els-cdn.com/

Hall, C. M. (2001). Trends in ocean and coastal tourism: the end of the last frontier? *Ocean & Coastal Management, 44*(92), 601–618. https://doi.org/10.1016/S0964-5691(01)00071-0

Hall, J. (2018 July 6). *8 incredible underwater hotels around the world*. Hotell Utter Inn, Vasterasfjarden, Vasteras. Retrieved November 28, 2022, from https://www.skyscanner.com/tips-and-inspiration/hotels/8-incredible-underwater-hotels-around-world & http://visitvasteras.se/aktor/hotell-utter-inn/

Kipreos, G., Konstantinakos, P., Anagnostopoulou, I., Perrea, A., & Kaprinis, S. (2016). Pavlopetri, the world's oldest submerged city: Analysis of dive tourism perspectives development in the region through S.W.O.T. analysis. *Hospitality & Tourism Management, 1*(3), 23–29. https://doi.org/10.11648/j.htm.20160103.11

Martínez Vázquez, R. M., Milán García, J., & De Pablo Valenciano, J. (2021). Analysis and trends of global research on nautical, maritime, and marine tourism. *Journal of Marine Science and Engineering, 9*(1), 93.

MED Sustainable Tourism Community, Enhancing a sustainable coastal and maritime tourism, Featured by BleuTourMed_C3. (n.d.-a). Retrieved November 28, 2022, from https://sustainable-tourism.interreg-med.eu and http://planbleu.org/sites/default/files/upload/files/BTM_Community_Brochure_18July2018.pdf

MED Sustainable Tourism Community. (n.d.-b). MED Sustainable Tourism Community contributes to the UNWTO Discussion Paper on "Sustainable Tourism for Development". Interreg Mediterranean; Sustainable Tourism. Retrieved November 28, 2022, from http://www.tourism4development2017.org and http://www.uni-med.net/wp-content/uploads/2017/07/UNWTO-Discussion-Paper-BleuTourMED_FINAL-.pdf

Mediterranean Snorkeling Report. (n.d.). Retrieved November 28, 2022, from http://www.snorkeling-report.com/region/snorkeling-mediterranean/

National Museum of Underwater Archaeology. (n.d.). A rich underwater archaeological heritage. Retrieved November 28, 2022, from https://www.spain.info/en/que-quieres/arte/museos/murcia/museo_nacional_de_arqueologia_subacuatica.html

NAUI. (n.d.). National Association of Underwater Instructors. Retrieved November 28, 2022, from https://www.naui.org

Nearly Zero Energy Hotels. (2013). Intelligent Energy Europe programme; Technical University of Crete. Retrieved November 28, 2022, from http://www.nezeh.eu/home/index.html

OECD. (n.d.). Regional Economy Data. Retrieved November 28, 2022, from https://stats.oecd.org/Index.aspx?DataSetCode=REGION_ECONOM. Year 2016, Real Growth Index of GDP 2007=100. See also Source: OECD. (2018). Real GDP forecast (indicator). https://doi.org/10.1787/1f84150b-en

Orams, M. (1998). *Marine tourism: Development, impacts, and management*. Routledge. ISBN 9781134771097.

Orams, M. B., & Lück, M. (2014). Coastal and marine tourism. In *The Wiley Blackwell companion to tourism* (pp. 479–489). https://doi.org/10.1002/9781118474648.ch38

PADI. (2018). The way the World Learns to Dive; The world's leading scuba diver training organization. Professional Association of Diving Instructors. Retrieved November 28, 2022, from https://www.padi.com/about/who-we-are

Padi Associates Environmental Consultants Ltd. (2010). *Master plan to support a sustainable diving industry for Malta.* San Gwann. Version I: December 2010; xi+55pp +I Appendix.

Parker, A. J. (1987, November). The Mediterranean, an underwater museum; Between 50 and 100 new sites located each year. *The Courier-UNESCO; Archaeology Under Water, 11*, 8–10. ISSN 0041-5278. Retrieved September 2, 2018, from http://unesdoc.unesco.org/images/0007/000765/076565eo.pdf

Piante, C., & Ody, D. (2014). *Blue growth in the Mediterranean Sea: The challenge of good environmental status.* MedTrends Project. WWF-France. 192 pages. Retrieved November 28, 2022, from http://d2ouvy59p0dg6k.cloudfront.net/downloads/medtrends_regional_report.pdf

Reportage for the visit of the famous French ocean explorer Zak Yb Kousto to Greece. The visit concerns the discovery of the lost Atlantis and other cultures in the area of Santorini Island. In this reportage documentary, the Greek Public TV (ERT) showed screens from the KALYPSO exploratory boat of KOUSTO, at the port of ZEAS. Retrieved January 8, 2023., from http://archive.ert.gr/15857/

Review of MuMart; Submarine trip and Underwater Museum of MuMart. (n.d.). TripAdvisor. Retrieved September 2, 2018, from https://www.lonelyplanet.com/italy/activities/golfo-aranci-mumart-museum-underwater-visit-in-submarine/a/pa-act/v-45093P41/359845

Romeo, N. (2015, November 3). Stunned archaeologists find 22 ancient Greek shipwrecks. *National Geographic.* Retrieved November 28, 2022, from https://news.nationalgeographic.com/2015/11/151103-greek-shipwreck-find-trading-route/

Sacchi. 2011. Analysis of economic activities in the Mediterranean: Fishery and aquaculture sectors. [pdf] Plan Bleu, Valbonne. Retrieved November 28, 2022, from http://planbleu.org/sites/default/files/upload/files/analyse_activites_econ_pecheEN.pdf

Shackley, M. (1998). 'Stingray City' – Managing the impact of underwater tourism in the Cayman Islands. *Journal of Sustainable Tourism, 6*(4), 328–338. https://doi.org/10.1080/09669589808667320
SSI. (n.d.). Scuba Schools International. Retrieved November 28, 2022, from https://www.divessi.com/el-IC/home/
Strati, A. (2006). Greece. In S. Dromgoole (Ed.), *The protection of the Underwater Cultural Heritage; National perspectives in light of the UNESCO Convention 2001* (2nd ed., pp. 97–126). Martinus Nijhoff Publishers. https://doi.org/10.1111/j.1095-9270.2007.140_1.x
Sustainable Development Goals (SDG 14): UN Western Europe. (2020) United Nations Western Europe. Retrieved January 16, 2023, from https://unric.org/en/sdg-14/
Tore Söderqvist, T., Soutukorva, A., Moreno-Arancibia, P., Liungman, O., Ahrensberg, N., & Paulsson, L. (2012). *Marine tourism and recreation in Sweden, a study for the economic and social analysis of the initial assessment of the MSFD*. Report produced for the Swedish Agency for Marine and Water Management. Havsoch vattenmyndigheten 2012-09-06. 2012: 2, ISBN 978-91-87025-02-0. Retrieved November 28, 2022, from http://www.tillvaxtbohuslan.se/turism/wp-content/uploads/sites/12/2014/04/marine-tourism-and-recreation-in-sweden-rapport-2012.pdf
UNESCO. (2001). The protection of the underwater cultural heritage. The benefit of the protection of underwater cultural heritage for sustainable growth, tourism, and urban development, elaborated by the UNESCO Secretariat and the Scientific and Technical Advisory Body of the Convention on the Protection of the Underwater Cultural Heritage (2001). Retrieved November 28, 2022, from http://www.unesco.org/new/fileadmin/MULTIMEDIA/HQ/CLT/pdf/UNESCO_UCH_Development_Study.pdf
UNWTO. (2017a). International year of sustainable tourism for development; A roadmap for celebrating together! World Tourism Organization 2016. Retrieved November 28, 2022, from http://cf.cdn.unwto.org/sites/all/files/intyear/unwto-brochure-celebrate-iy20172.pdf
UNWTO. (2017b). International year of sustainable tourism for development; Why tourism? (-Five Pillars). Retrieved November 28, 2022, from http://www.tourism4development2017.org
UNWTO. (2018). *Tourism for development volume II: Good practices*. World Tourism Organization. Retrieved November 28, 2022, from https://www.e--unwto.org/doi/pdf/10.18111/9789284419746

Uyarra, M. C., Watkinson, A. R., & Côté, I. M. (2009). Managing dive tourism for the sustainable use of coral reefs: Validating diver perceptions of attractive site features. *Environmental Management, 43*, 1–16. https://doi.org/10.1007/s00267-008-9198-z

VDST. (2023). Association of German scuba divers. Retrieved October 28, 2023, from http://www.vdst.de

Viator-A Trip Advisor Company. (2023). Top Europe submarine tours. Retrieved October 28, 2023, from https://www.viator.com/Europe-tours/Submarine-Tours/d6-g3-c67

13

Proximity Tourism as a Sustainable and Resilient Strategy: Building Back Better for Local Communities and Destinations

Nina Balan and Stefania Escobar

Introduction

An increased awareness of climate change, the environmental crisis and the aftermath of the COVID-19 pandemic contributed to an increasing attention towards proximity tourism as a strategy for a sustainable recovery of tourism destinations. However, despite the relevance of the topic, the existing literature in tourism and leisure studies on proximity tourism is still scarce (Salmela et al., 2021). In this chapter, we explore a case of proximity tourism that positively contributed to the sustainable development of a destination during a time of crisis.

Proximity tourism has been defined as a "form of tourism that aims to (re)discover the places nearby, those that, precisely because they are too

N. Balan (✉)
University of New Orleans, New Orleans, LA, USA

S. Escobar
Free University of Bozen-Bolzano, Bolzano, Italy

N. Stylos et al. (eds.), *Contemporary Marketing Management for Tourism and Hospitality*, https://doi.org/10.1007/978-3-031-65049-9_13

close and familiar, are not considered as possible destinations for holidays or weekends off" (Bieliková & Palenčíková, 2021).

The concept of proximity challenges the traditional way of defining tourism based on nearness and farness or through the interplay of the ordinary and extraordinary (Urry, 2002). In this chapter we explore a case of proximity tourism that allow Lunigiana, a rural destination in the Tuscany region in Italy, that is aiming towards a new concept of tourism and detach from the strategies pursued by major destinations in the area giving priority to its citizens, its culture and the future of its natural environment and resources with positive feedback from tourists and locals. Still, the destination is at an embryonal stage of development and therefore additional steps are needed in order to ensure its promising sustainable development in the long-term.

Proximity Tourism During Crisis

Given current climate crises and challenges, proximity tourism risks to be considered a temporary practice while waiting for international travel to recover and return to pre-crisis habits and frequency of traveling. However, proximity tourism has the potential to become a long-term strategy at the destination level that could be sustainable for the hosting communities, implementable at reduced costs and replicable for destinations struggling with similar challenges (Govers et al., 2008). This reasoning follows the latest claim advocated by previous studies for a change after crisis times to build back better through sustainable tourism that can reconstruct the tourism system with a focus on the local community and destination preservation rather than returning to pre-pandemic unsustainable tourism evolution that neglects the impact of the industry on destination resources and inhabitants (Cheer et al., 2021; Higgins-Desbiolles, 2020; Sigala, 2021).

Although the tourism literature concerning the development of new tourism services and marketing strategies has focused prominently on external tourism flows, the current proximity-oriented literature has started to acknowledge the possibilities related to the development of tourism services and marketing for local and intraregional tourists (Salmela et al., 2021). Studies show that proximity tourism can work as a strategy to tackle some prominent issues in the field, including seasonality and disregard for local heritage (Diaz-Soria & Llurdés Coit, 2013). Thus, as previously stated by Diaz-Soria and Llurdés Coit: "Tourism supply and demand are being deeply transformed, and territories need to improve their adaptability in order to respond properly to new tourist demands" (2013). Changes in the economy, such as the economic and financial crises in 2008–2009, Severe Acute Respiratory Syndrome (hereinafter, SARS) in 2003 or COVID in 2019, turned tourism marketers' gaze towards proximity tourism as a possible alternative for international tourism, as it is a more affordable and equally accessible way of traveling between different areas in the same region (Bertacchini et al., 2021). This is the case of Lunigiana, which has enhanced its territorial treasures and local activities to promote local tourism based on natural environments such as densely forested terrain, historic towns and "agriturismi", most of which remain, blissfully, under the radar of mass tourism (Mackay, 2019).

From a spatial point of view, proximity tourism has mainly been studied as a recent phenomenon, related first of all to the increase of environmental awareness and second of all, as a mechanism of adaptation in a context of economic crisis. This is when traveling far is not possible anymore; tourists would choose closer destinations. An example is the drastic decrease in international travel to closing borders restrictions dictated by the global COVID-19 pandemic, which gave new light on the concept of proximity tourism and its meaning. The concept is also notorious due to increased awareness of the impacts of climate change and the urgency of our environmental crisis. Since adopting the Paris Agreement in 2015, shorter distances and lower carbon means of transportation have been encouraged to reduce carbon footprint. Hence, among many definitions given to proximity tourism, one refers to a form of tourism focusing on close local destinations that involve re- discovering sociocultural activities and re-starting the economy in rural parts of the cities. Shifting focus on proximity tourism and reorienting outbound tourism can be a profitable strategy to help destinations overcome crises when inbound tourism falls to zero.

One strategy to shift the interest in local manufacturing and workforce is the investment in cultural heritage (and other forms of culture) which often are claimed to be beneficial for a local economy, not only in terms of cultural consumption but also in the form of increased employment and income (Bowitz & Ibenholt, 2009). For instance, small-scale manufacturing businesses, handicrafts, and local artisans often offer work opportunities regardless of education level and background. Artisans and small-scale businesses often produce goods by hand or with small tools. They are often full-time producers and generally sole proprietor's businesses or have a couple of employees, and they sell goods online (a trend predominantly popular since the COVID-19 pandemic) or at the local market on the weekends. Small-scale business owners and self-employed artisans not only carve out a living in the hospitality sector but also enhance local signature dishes and embrace the peculiarity of unique traditions, making them memorable for visitors. It is well known that small-scale businesses constitute a substantial part of the tourism industry, reflecting the prominence of such businesses in developing economies at large (Dahles & Susilowati, 2015).

Proximity Tourism in Lunigiana

In this chapter, we investigate the proximity tourism phenomenon and its repercussions on sustainable tourism development through the lenses of a European destination that re-evaluated the way tourism has been practiced for years and adopted new-tech-based tools to enhance local entrepreneurship and promoted local goods and products throughout times when most of the small scale businesses chose to close down due to temporary crises. We interviewed different stakeholders, such as an app innovator and developer, the president and project manager of Lunigiana destination management organization (DMO), small-scale business owners and local tour guides. In addition, we consulted tourism statistics from open data platforms from the Tuscany region, newspaper articles, destination reports and websites, data from the app and scientific articles (Diaz-Soria, 2017; Jeuring & Diaz-Soria, 2017).

Lunigiana Destination Characteristics

The Lunigiana (pronounced [luniˈdʒaːna]) is a historical territory of Italy, which today falls within the provinces of Massa Carrara, Tuscany, and La Spezia, Liguria (Terre del Magra, 2016). Lunigiana is located 20 km from the sea, and 30 km from the ski resorts, making it a favorable location for summer and winter sports. The region is characterized by a pre-mountainous geography with valleys, mountains, streams, waterfalls and woods where medieval castles and small stone villages dot the immense territory, which today is divided into 14 municipalities enclosed within the Tourist Area of Lunigiana (Fig. 13.1).

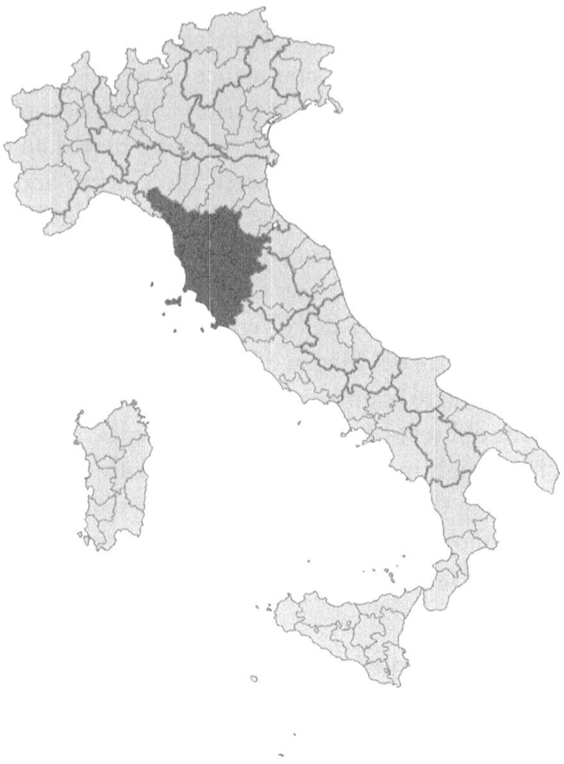

Fig. 13.1 A map of Lunigiana, Tuscany. (Source: Wikicommons, Public Domain, 2008)

Lunigiana is a perfect place to escape from overcrowdedness of neighboring tourist destinations (such as Florence and Cinque Terre) and immerse in an uncontaminated green and slow environment. It is a purely naturalistic territory waiting to be discovered, with little population and many animals in their natural environment, where you can get lost in the chestnut groves, truffle hunts, walk in the ancient paths, visit small villages still inhabited and take a wild swim in the hundreds of waterfalls and pools (Wiki Commons, 2014).

Lunigiana's historical roots go back to the second century BC during the Roman Empire. The name comes from the city of Luni, a famous ancient seaport of the river Magra (Terre di Lunigiana, 2022).

Lunigiana is the land of a hundred castles, imposing buildings and fascinating ruins. This extensive parkland is well regarded by locals for its verdant valleys and picturesque Borghi (Mackay, 2019). One of the particularities of Lunigiana is its abundance of semi-dilapidated castles (Fig. 13.2). This was once a contested territory. It was occupied by French and Spanish forces for several hundred years and played an essential role in the brutal and protracted war between the Guelph and Ghibellines. There are said to have been as many as 160 fortresses in the vicinity during the middle ages, and 30 remain to this day. The most important include the Castello Malaspina, near Carrara, the Castello di Castiglione del Terziere, and the Fortezza della Verrucola, which is now home to a printing museum (Terre del Magra, 2016).

The region is home to countless tiny villages many of which have seen their populations obliterated in the past few decades. During World War II, there was heavy fighting, and Nazi forces virtually extinguished several communities during the liberation. Tourists who visit the region are prepared for a gastronomic gem or to follow in the footsteps of medieval pilgrims along the Via Francigena.

Revitalization of Lunigiana

During times when destinations were suffering from a slow recovery of tourism numbers aggravated by a still high risk perception, Lunigiana managed to keep tourists interested throughout the months after the

Fig. 13.2 Lunigiana semi-dilapidated castles. (Source: Wiki Commons, 2014)

COVID-19 pandemic was regulated. The success is likely due to pre-pandemic efforts to brand the destination as a unified entity giving new value to natural attractions and the development of initiatives based on smart technologies that benefit the environment and allow visitors to explore new territories. Most importantly, the final boost was given by strategies developed during the pandemic confinement. According to interviewee #4, staycation was already a national trend in Italy when the pandemic started, and Lunigiana being positioned on advantageous sprawling land, travelers escaped big crowded cities to access more space. Thus, guests naturally chose territories like Lunigiana throughout the pandemic times. This need for nature and space caused an influx of tourists.

The most famous smart initiative was the geo-localization of attractions in natural and rural areas (Stylos et al., 2021a, 2021b) connected with an application called *Lunigiana World* developed for tourists and residents alike. The strategy was initiated in 2018. As an outcome, the app was developed, which geo-localized all-natural attractions in Lunigiana and shared them through social media, namely Facebook and Instagram reaching a number of more than for thousand downloads with a predominant interest towards natural activities from users according to interviewee #3. Concurrently, local governments started to develop partnerships to initiate a common branding of the destination, and this is how a number of entrepreneurs became part of the application. The efforts to create a smart initiative tool served as both an analysis of what had happened in the crisis period and a recommendation for what should happen in response (Stylos, 2020, 2022a, 2022b).

The app is supplied with several features available in English and Italian, including places to visit, historic hamlets, lodging and hospitality, food and beverage options, events to attend, sports, traditions during the holiday months leading to the Ferragosto celebration, weather forecast, info point, and mountain biker (MTB) trail. The app also features local products such as grain, apples, honey, beer, olive oil, truffle, pasta, and chestnuts. Tourists are often interested in visiting local farmer's markets and supporting local businesses while experiencing a unique cuisine particular only to Lunigiana. Each sub-section on the app is accompanied by a short description of the history and story of the founding family, in addition to images to magnify the tourist's interest in local agriculture.

The app further contributed to the development of a common vision of tourism in Lunigiana spearheaded by SIGERIC and Visit Lunigiana. SIGERIC is a cooperative of tourism professionals and tour guides of Lunigiana, which plays the role of a destination management company (DMC) and DMO for the region. The aim of SIGERIC is to give the opportunity to any traveler (tourists or else) to discover Lunigiana with a special focus on historical, cultural and natural experiences that allow us to keep ancient traditions and prime local products (SIGERIC, 2015). This common vision allowed for single municipalities to join in a common effort shifting from strategies focused on their single territories into a vision of destination that gathered their attractions and businesses under the Lunigiana name, gaining more visibility and traction at a national and international level.

Meanwhile, *Visit Lunigiana* is a local association that aims to develop initiatives to promote the territory, with particular attention to enhancing local products and ancient traditions and helping all tourism businesses to work as part of a coordinated network (Visit Lunigiana, 2023). This unified vision and effort made Lunigiana an outlier when other regional destinations in Italy and cities worldwide were trying to recover from an almost total halt in national and international arrivals. In May 2020, when Italy stopped the quarantine restrictions, Lunigiana saw an immediate increase in tourist presence from the provinces and neighboring regions. Given the imminent return of tourism after the Italian Government announced the official date of Confinement end (May 4, 2020, Corriere della Sera, 2020), the local DMC, with the support of local administrations and associations on the territory developed in ten days a tourism strategy for Lunigiana supported by municipalities, associations of the territory (Lunigiana.it, 2020). This strategy, together with the visibility of the app, created the perfect mix for the success of Lunigiana tourism in a time of crisis, which became one of the very few areas in the Tuscany region that kept tourism flows (Il Tirreno, 2021; La Voce Apuana, 2021).

Aware of this advantage, in 2020 Lunigiana strategically unified its brand with the following focus: nature-based tourism, proximity tourism, deseasonalization, and remote working, all under the common aim of developing sustainable tourism as an enriching tool for community development (SIGERIC, 2020). After confinement due to COVID-19, the risk perception was still high, and the risk of infection was still present among tourists (as stressed by all interviewees in this study). The opportunity to enjoy a destination without worrying about these two factors allowed Lunigiana to be favored amongst famous destinations nearby such as Pisa, Firenze or Cinque Terre in Liguria. According to interviewee #5, from a tourism point of view COVID was positive as it made Lunigiana visible to a huge number of tourists that normally went towards nearby seaside destinations.

Indeed, Lunigiana is located 30 km away from the seaside and therefore the flow of tourists normally passing through for the seaside, after COVID-19 confinement started to make stops in Lunigiana. Due to the pandemic restrictions, DMOs saw a new pattern in traveling behavior for

Italians who planned seaside trips, stopping in the Lunigiana region and rediscovering territories that once were considered 'under tourism' areas (Lunigiana.it, 2020).

According to the data, in 2020, there was a predominant presence and arrivals of Italian visitors compared to international visitors. This is directly related to the increasing number of COVID-19 cases and the federal restrictions, which continued after the mandatory quarantine at the beginning of 2020. Although the year 2020 numbers show a decrease in tourism compared to the year 2019, considering that rigorous policies were enforced regarding traveling and leaving homes during five months of the year, with March to May as the first restriction period. October to December was second, creating almost six months of quarantine (Openpolis.it, 2023). Despite restrictions, the interest in visiting Lunigiana is noticeably high, with increased interest and visits from nearby regions.

It has often been argued that the tourism sector has high resilience and the capacity to adapt and recover from catastrophic or unexpected phenomena (Romagosa, 2020). Recent studies argue that once the worst moments have passed, it will gradually return to a certain level of normality, or at least to a situation similar to pre-crisis (Ferrari et al., 2022). Nevertheless, the return often requires building greater resilience. Resilience is the capacity of any entity—an individual, a community, an organization, or a natural system to prepare for disruptions, recover from shocks and stresses, and adapt and grow from a disruptive experience. Developing resilience also allows us to bounce back from a crisis, learn from it, and achieve revitalization (Rodin, 2014). Ideally, once a destination becomes more adept at managing disruption and skilled at resistance building, it could learn to take advantage of new opportunities in good and bad times. Rodin calls this concept a resilience dividend; when the return after a disruption is achieved, it is about achieving a significant transformation that yields benefits even when disruptions are not occurring (2014, p. 4). The worldwide examples post-COVID pandemic are admirable. Cities saw new energy flow back into their downtown areas and neighborhoods' main streets. Several cities in countries like Poland, France, Italy, and the United States were blocking off streets allowing children to play outside safely and ride bikes. Small businesses got creative with organizing events and outdoor dining tables as a form of social connection after months of isolation and gathering the community. Residents have started to spend more time in public green spaces that have become very important to them, allowing them to walk without masks, meet with friends and use outdoor gyms and shared spaces. Due to population density and the complexity of human interactions, cities experience specific problems arising from their functioning in the era of the COVID-19 pandemic. Many scientists who published on the post-pandemic crisis agreed on the need to shape pandemic resilient cities by making changes to the functional-spatial structure and preparing cities for rapid revitalization [regeneration] in the post-pandemic period (Majewska et al., 2022).

In the United States, a number of small and medium-sized communities, such as Greenville (South Carolina), are growing in population and attracting business because they started to invest in their people and places differently (Preuss, 2021). Thrillist Travel (2018) compiled some of the best small towns in the United States to move to before they become too popular. For Lunigiana, one of the strategies to practice resilience after crises was enhancing proximity tourism, especially when traveling across regions was still not allowed or gradually restarting (Il Tirreno, 2021). To this end, the application became extremely useful to outcast the different attractions of Lunigiana together with a strategy developed by local associations to promote the destination through different channels directly to consumers and the local community.

According to the feedback received, the owner of the application (interviewee #4) confirmed that through the activation of the smart app, the inhabitants of Lunigiana actually rediscovered their territory that once was an element of shame given that there is an ongoing comparison with nearby giant destinations like Florence. Indeed, although the app is independent of destination strategies and does not receive public funding, it had a crucial role in the redefinition of the destination and its tourism boost starting from its proximity tourism.

Lunigiana, as a destination, offered great green valleys without overcrowdedness. An offer for tourist attractions based on rural and natural activities and an image of an under-tourism destination, less populated with visitors than most parts of the Tuscany region. These elements shifted from being a negative feature or an element that prevented

Lunigiana from being competitive in the region to a positive feature that helped Lunigiana gain competitiveness organically (Lunigiana.it, 2020).

Lunigiana's strategy allowed to rediscover the value of nature and open air activities not only for tourists outside the region but also and most importantly for locals (Il Tirreno, 2021). Locals were able to rediscover the value of their territory in terms of natural attractions, gastronomy and resources. This aspect was fundamental in Lunigiana's tourism strategy as stated by DMO's project manager, (interviewee #1) explaining how proximity tourism is fundamental for Lunigiana, and how locals are the best ambassadors of the territory and therefore it is crucial to cherish their presence and activities. A local can completely change the experience that a tourist has on a destination.

The inclusion of local communities "exerts a powerful influence on the tourists' experience". They can provide useful hints for tourists on a variety of tourism products, becoming a true ambassador for the territory through word of mouth, web reviews, participation and exchange of opinions. Through the development of proximity tourism, locals were empowered to acknowledge the value of the territory's resources and how to preserve them. Indeed, after the development and growth of the app *Lunigiana World*, many tour guides, videos and books were developed around Lunigiana history and landscape, and translated in different languages for the visitors.

In their effort to improve the involvement of local businesses and the community at large, SIGERIC offers during low season special price access, offers and discounts to all tours for locals, including e-bike services and other activities to gain locals' feedback allowing the local community to be actively involved in the destination development process, and share their experiences with tourists either through online reviews or word of mouth.

However, Lunigiana has also seen examples of areas where locals were not involved. Based on the conversations with interviewees, we learned that there are waterfalls, priorly deserted, that during the pandemic had in between 100 and 200 people at the time, and some locals argue that this presented as a problem. This proves that if residents are not involved at all stages of planning, permitting and policy-making process in a tourism strategy development, they can negatively influence the tourist experience, and moreover the community participation becomes an obstacle instead of an opportunity. Therefore, strategic planning is needed to preserve the areas that are seeing a rapid change in terms of tourism presence and equip them to confront these new increasing numbers.

Through the example of Lunigiana, we found out that proximity tourism may serve as a tool for destinations to rethink tourism development in several ways: (i) boost the local economy; (ii) increase destination resilience to environmental hazards; (iii) rediscover places nearby and local heritage for residents and visitors alike, (iv) shed light on local resources and support small and medium-sized enterprises (SMEs), (v) enhance new traditions in host communities, all while decoupling the number of carbon emissions derived from traveling further. However, it is important to notice that a key role in the successful implementation of the recommendation above remains the participation of the local community (Vassiliadis & Stylos, 2022).

Nature-Based Tourism

Among the most popular tourism activities to explore the Lunigiana territory remains nature based attractions such as bike paths, water activities, climbing, waterfalls and hiking. According to Kuenzi and McNeely (2008), more and more people are living an urban life, and the amenities and conveniences that come with globalization are leading to a near-complete disconnect from nature. has the opportunity to offer unique ways to connect in nature by enhancing local products during its peak seasons while in other areas, the seasonality of fruit and vegetables virtually has ceased to exist. For example, autumnal visits are rewarded with fresh, intensely scented porcini mushrooms that sprout under chestnut trees in fecund woods and hills (James et al., 2017). Wild herbs cover fields, and 5000 scattered hives produce the region's famous chestnut and acacia honey. These fruits of the forest and other regional delicacies, including Zeri lamb, freshly baked *focaccette*, crisp and sweet *rotella* apples, boiled pork shoulder, *caciotta* (a delicate cow's-milk cheese), *bigliolo* beans, local olive oil and Colli di Luni wines, can be found in Lunigiana (Lonely Planet, 2022).

Seasonality

The cuisine plays an important role in the success of the Lunigiana region. Thanks to its geography, on the boundary of Tuscany, Liguria and Emilia-Romagna, the area is well positioned to take elements from each of these culinary traditions (Mackay, 2019). The value of natural products such as porcini mushrooms, truffles, and chestnuts always present in Lunigiana forests became an attraction with both organized tours to retrieve these products with local guides and local restaurants and shops, allowing the continuity of the season until November. Local small scale business owner interviewee 5 cultivates, sells and manufactures/produces preserved truffles since 2020 to local and international buyers. In addition to being an employer to seven employees, the owner is a tourist guide, truffle hunter and a guide for excursions and site visits in the region. Therefore, this is a great example of SMEs' holistic contribution to socio-economic development and seasonality of the place. Seasonality represents one of the most important issues of tourism demand and it has an extremely significant impact on many aspects of tourism development (Zvaigzne et al., 2022).

The preservation of local traditional activities allows on the one hand to keep those activities alive from generation to generation but also allows for developing new tourism products in the area. Indeed, new activities such as guided tours to rediscover local products growing in the forests nearby such as truffles, mushrooms, chestnuts cooking classes (Fig. 13.3), and tastings are organized in the area especially during October and November to showcase these traditions to tourists but also during low season through December to March (Lunigiana World, 2022; Visit Lunigiana, 2022). In addition, the development of festivals in low season helped promote local sustainable products and revitalize the territory and local businesses such as the Truffle Festival of Bagnone held in December 2022 that involved the destination as a whole promoting local products and local traditions (Proloco Bagnone, 2022). As an outcome, the season fully extended throughout October and November which provides increased value for community development where tourism products based on traditional activities of a local community play a main role (as explained by interviewee #1 and #4).

Fig. 13.3 Chestnut manual processing. (Source: Wiki Commons, Public Domain, 2012)

Remote Working

During the pandemic, the abrupt closure of many offices and workplaces ushered in a new era of remote work for millions of employees worldwide. Communities like Pontremoli, Tuscany have swiftly learned to adapt to the new reality and offered remote workstations with shared services in geographically beautiful areas and architectural spaces decorated with paintings by local artists and located inside a former monastery, with rooms for meetings and phone calls, cafeteria and courtyards with beautiful lawns. It is just a few steps away from the city's many bars, restaurants and pizzerias for easy access within walking distance. Revitalization of local communities has been practiced by offering unrented properties for rent to remote working individuals. The initiative had a strict policy requiring local municipalities to invest in an appropriate internet connection and other amenities, including common areas for remote working in small villages, to be eligible for rent. An essential element of this strategy consists of the criteria defined to access these properties. Although tourists could still apply for this rental property, the application clearly stated that only those willing to stay for a certain period and be involved in the village activities were welcome. In this way, tourists became an enriching part of the community.

Practicing Sustainable Tourism

According to local policymakers, Lunigiana is not destined and will never reach the number of the neighboring popular tourist destinations such as Florence or Pisa. This seems to be why a plan to assess the sustainability of tourism on the territory's resources and infrastructures is still at an embryonal phase of development. Yet, it is crucial to address this strategic planning with a focus on sustainability, as climate change has been enforcing the need for climate resilience and adaptation pathways to preserve our natural environment. Thus, as Lunigiana plans and prepares for more socio-economic-driven initiatives, sustainability must lie at the core of all new initiatives, whether promoting a new territory or maintaining existing popular sites and regions.

Sustainability tells us to look for balances between three dimensions: the environment, society and the economy (Romagosa, 2020). Tourism can be a good tool for the region's socio- economic development if balance is successfully achieved and tourism benefits all three dimensions. Thus, it will be necessary to find a social balance in terms of equity and justice, as well as an economic and environmental one, in every destination. This is where tourism planning and management policies come into play in implementing sustainability and resilience at all scales (local, national and international) and with appropriate forms of governance, integrating the public and private sectors in a coordinated manner (Romagosa, 2020). If sustainability is implemented with those values in mind and supported by locals and visitors, chances for a destination to thrive are very high. Many destination management organizations struggle with implementing sustainability due to the competitive market with short-term rentals and tour operators that have for years promoted popular destinations at a very high level. A lack of policy implementation and communication between municipalities and entities that enhance local sustainability and local market-driven entities leads towards mass tourism and over-tourism concepts that later become unsustainable for the territory. Lunigiana shows examples of sustainability in many of their strategies. Lunigiana is promoting local products deeply rooted in the heritage of the territory as an added experience for tourists allowing these products to survive through the years and making locals more aware of the power of local heritage preservation as an enriching element for community development.

Remote working option represents another important sustainability milestone of the Lunigiana region. Through this high tech project, the aim is to revitalize rural centers by having semi-abandoned buildings repurposed for remote workers by offering them several services, benefits, connection with nature and making workers part of the community. This project contributes to community empowerment but also to the development of new businesses, services and activities with an influence on both economic and social sustainability. Embracing innovation while preserving history and authenticity is a key role for sustainability. Finally, the focus on proximity tourism allowed locals to rediscover their territory, become aware of the valuable resources it has to offer, but also allow for lower greenhouse gas emissions given from mobility outside the region which contributes to social, economic and environmental sustainability.

Sustainable Tourism for Receiving Communities

There is a gap in understanding the benefits of implementing sustainability at all scales. We learned through local knowledge and expertise that the work is arduous to implement due to a lack of support for sustainable destination management practices. Those who do not believe in the preparation of tourist sites become a barrier. Not governing the influx of tourists is non-beneficial to the sites and the receiving community, and the lack of such regulations does not bring income or any benefits.

It is indeed hard to shift away from a focus that prioritize economic growth and instead move towards the concept of sustainable tourism; according to interviewee #4, tourism numbers are greatly improving, but the receiving community and sites need to improve as well. Results will lack if there is a lack of planning, management and data monitoring.

Considering the impact of tourism on resources, infrastructures and community are essential elements for any destination in the process of developing tourism products, attractions and experiences as authenticity loss, resident negative tourist perception can negatively influence the success of a destination management strategy (Jiménez-Barreto et al., 2020). To this end, gathering tourism data becomes crucial to understand the tourism system in terms of destination features but also to understand the limits of a territory and its accessibility (Kolotouchkina et al., 2022). Differently from its big neighboring destinations such as Florence, Pisa and Cinque Terre, Lunigiana claims to have from the very beginning put sustainability at its core, and some of the strategies show clearly this direction. If Lunigiana's mindset to start a tourism strategy rooted in sustainable principles is admirable, more efforts are needed to boost the sustainable development of the destination in terms of protection of the local community, resources and infrastructures from tourism impact.

Even though a destination is located at the initial phase of tourism development compared to its neighbors, in order not to reach that level of tourism impact, a planned strategy outlining possible impacts and measurement of those impacts is essential. In addition, receiving communities need to be involved to understand possible issues that may arise from tourism activities in certain areas as well as to get feedback on actions taken by municipalities.

Conclusions and Recommendations

This section will provide conclusions and recommendations for destinations aiming to start or restart their tourism development strategy with sustainability at its core. Lunigiana represents an outlier tourism destination due to its favorable conditions, self-distancing and low-risk perception of open-air activities, and reputation as an under-tourism destination. Thus, the destination could register positive tourism numbers when destinations gradually recover from the total halt of national and international tourism arrivals. At the core of their strategy, we find proximity tourism, the rediscovery of the territory for locals, and the value of local traditions, nature-based activities, and village revitalization. Thanks to the interviews conducted and data gathering, we also learned that often elderly are not as enthusiastic about the tourism drive compared to the younger population, who are more equipped to implement innovative strategies to re-discover old traditions and customs. Nevertheless, for continuing socio-economic growth and change, adaptation to the new reality is required; thus, Lunigiana finds the perfect balance between offering socio- environmental sustainability for the elderly, preserving culture and traditions while engaging the youth, and more progressive-thinking residents embedding innovative tourism activities. The strategic planning through smart technology implementation, persistent promotion, and adaptability put into place by stakeholders, municipality leadership, and the destination management organization (DMO) proved successful during the crises and a great example of resilience for the future.

We will proceed with three recommendations that could help DMOs and organizations alike aiming to develop similarly:

- Proximity tourism is a valuable tool as it allows the inhabitants to rediscover the value of the territory and the value of tourism for community development. It is also a great strategy to involve residents in tourism planning. From our interviewees, it became relevant to highlight the importance of investing in tourism with great regard and 'respect' for nature and local heritage.
- Vision setting was the second most recommended strategy: Interviewee #4 claims that "without a vision, fundings are useless", which refers to the vital relevance of developing sustainable long-term tourism planning that considers, from the planning phase, all possible impacts of tourism on a destination's resources, infrastructures and the local population.
- In addition, it is crucial to consider the importance of data: data can help understand the impact of tourism on a destination by measuring tourism flows, peak seasons, and tourist current and future preferences (Gallego & Font, 2021; Fromel et al., 2020). Therefore, a destination

needs to retrieve data, analyze it and develop evidence-based strategies accordingly. Given the vast amount of data that the current running applications have to offer, an essential and smart strategy would be to partner with these applications to understand the data and how to analyze it and afterwards develop together a destination management strategy that considers the impact of tourism on the territory and makes tourism a tool for community development.
- Finally, we acknowledge that a tourism development strategy based on proximity tourism is not applicable in every kind of destinations such as small islands destinations. However, some of the elements of this strategy could be inspirational for any kind of destinations as it is based on sustainability principles that are replicable for any territory.

Recommended Further Reading

Carrubbo, L. (2010). *Service science, management, engineering and design and its suggestions for destination management.* Proceedings of 13th Convegno TOULON.

Ferrari, G., Bocci, E., De Santis, C. D., & Rombai, L. (2022). Tracce di Resilienza in Tempo di Pandemia lungo itnerari storici in Lunigiana e Tuscia. *Turismo e Psicologia.*

Lemmi, E., & Deri, M. G. (2020). A new model for the tourism renaissance: The case study of the Tuscan village of San Pellegrino in Alpe.

Meneghello, S. (2023). Mapping tourist landscapes in pandemic times: A dwelling-in-motion perspective. *Tourism Geographies*, 1–16.

Stylos, N., Koroneos, C., Roset, J., González-Sánchez, C., Xydis, G., & Muñoz, F. S. (2018). Exergy as an indicator for enhancing evaluation of environmental management performance in the hospitality industry. *Journal of Cleaner Production, 198*, 1503–1514.

Stylos, N., & Vassiliadis, C. (2015). Differences in sustainable management between four-and five-star hotels regarding the perceptions of three-pillar sustainability. *Journal of Hospitality Marketing & Management, 24*(8), 791-825.

References

Bertacchini, E., Nuccio, M., & Durio, A. (2021). Proximity tourism and cultural amenities: Evidence from a regional museum card. *Tourism Economics, 27*(1), 187–204.

Bieliková, H., & Palenčíková, Z. (2021). Sustainable tourism development in times of the COVID-19 pandemic: Case study of Slovakia. In *Handbook of research on the impacts and implications of COVID-19 on the tourism industry* (pp. 823–842). IGI Global.

Bowitz, E., & Ibenholt, K. (2009). Economic impacts of cultural heritage – Research and perspectives. *Journal of Cultural Heritage, 10*(1), 1–8.

Cheer, J. M., Lapointe, D., Mostafanezhad, M., & Jamal, T. (2021). Global tourism in crisis: Conceptual frameworks for research and practice. *Journal of Tourism Futures, 7*(3), 278–294.

Corriere della Sera. (2020). Coronavirus, la fase 2: le condizioni per le Regioni che vogliono partire prima del 4 maggio. https://www.corriere.it/cronache/20_aprile_20/coronavirus-fase-2-condizioni-le-regioni-che-vogliono-partire-prima-2a328f02-82bc-11ea-86b3-8aab0c7cf936.shtml

Dahles, H., & Susilowati, T. P. (2015). Business resilience in times of growth and crisis. *Annals of Tourism Research, 51*, 34–50.

Diaz-Soria, I. (2017). Being a tourist as a chosen experience in a proximity destination. *Tourism Geographies, 19*(1), 96–117.

Diaz-Soria, I., & Llurdés Coit, J. C. (2013). Thoughts about proximity tourism as a strategy for local development. *Cuadernos De Turismo, 32*, 303–305.

Ferrari, G., Bocci, E., De Santis, C. D., & Rombai, L. (2022). Tracce di resilienza in tempo di pandemia lungo itinerari storici in Lunigiana e Tuscia. *Turismo e Psicologia*.

Fromel, K., Kudlacek, M., & Groffik, D. (2020). Tourism and physical activity preferences: Development and sustainability strategy. *Sustainability, 12*(21), 8824.

Gallego, I., & Font, X. (2021). Changes in air passenger demand as a result of the COVID-19 crisis: Using Big Data to inform tourism policy. *Journal of Sustainable Tourism, 29*(9), 1470–1489.

Govers, R., Van Hecke, E., & Cabus, P. (2008). Delineating tourism: Defining the usual environment. *Annals of Tourism Research, 35*(4), 1053–1073.

Higgins-Desbiolles, F. (2020). Socialising tourism for social and ecological justice after COVID-19. *Tourism Geographies, 22*(3), 610–623.

Il Tirreno. (2021). Così la Lunigiana ha tenuto a galla il turismo toscano ai tempi del Covid. https://www.iltirreno.it/massa/cronaca/2021/09/24/news/cosi-la-lunigiana-ha-tenuto-a-galla-il-turismo-toscano-ai-tempi-del-covid-1.40739105

James, A., Ravichandran, S., Chuang, N. K., & Bolden, E., III. (2017). Using lifestyle analysis to develop lodging packages for staycation travelers: An exploratory study. *Journal of Quality Assurance in Hospitality & Tourism, 18*(4), 387–415.

Jeuring, J. H. G., & Diaz-Soria, I. (2017). Introduction: Proximity and intraregional aspects of tourism. *Tourism Geographies, 19*(1), 4–8.

Jiménez-Barreto, J., Rubio, N., & Campo, S. (2020). Destination brand authenticity: What an experiential simulacrum! A multigroup analysis of its antecedents and outcomes through official online platforms. *Tourism Management, 77*, 104022.

Kolotouchkina, O., Barroso, C. L., & Sánchez, J. L. M. (2022). Smart cities, the digital divide, and people with disabilities. *Cities, 123*, 103613.

Kuenzi, C., & McNeely, J. (2008). Nature-based tourism. In O. Renn & K. D. Walker (Eds.), *Global risk governance: Concept and practice using the IRGC framework* (pp. 155–178). Springer.

La Voce Apuana. (2021). Turismo apuano: i livelli pre-covid sono lontani. Bene Massa e Lunigiana, male Carrara. https://www.voceapuana.com/economia/2021/12/17/turismo-apuano-i-livelli-pre-covid-sono-lontani-bene-massa-e-lunigiana-male-carrara/50355/

Lonely Planet. (2022). The Lunigiana. https://www.lonelyplanet.com/italy/tuscany/the-lunigiana

Lunigiana World. (2022). Escursioni – Ricerca del Tartufo. https://www.lunigianaworld.it/ricerca-tartufo/

Lunigiana.it. (2020). Turismo, La Lunigiana si promuovo post Covid. https://www.quinewslunigiana.it/fivizzano-la-lunigiana-si-promuove-sui-social-media.htm

Mackay, J. (2019). Discovering the Lunigiana: Tuscany's hidden gem. Retrieved December 27, 2022, from https://www.italymagazine.com/featured-story/discovering-lunigiana-tuscanys-hidden-gem

Majewska, A., Denis, M., Jarecka-Bidzińska, E., Jaroszewicz, J., & Krupowicz, W. (2022). Pandemic resilient cities: Possibilities of repairing Polish towns and cities during COVID-19 pandemic. *Land Use Policy, 113*, 105904.

Openpolis.it. (2023). Coronavirus, l'elenco completo degli atti. https://www.openpolis.it/coronavirus-lelenco-completo-degli-atti/#lo-statodi-emergenza-nel-nostro-paese-e-durato-26-mesi-cio-e-servito-per-garantire-loperativita-della-struttura-creata-per-gestire-la-pandemia

Preuss, I. (2021). *Recast your city: How to save your downtown with small-scale manufacturing*. Island Press.

Proloco Bagnone. (2022). 1 Festival del tartufo in Lunigiana. https://www.apt-massacarrara.it/wp-content/uploads/2022/12/Programma-del-1-Festival-del-Tartufo.pdf

Rodin, J. (2014). *The resilience dividend: Being strong in a world where things go wrong*. Public Affairs.

Romagosa, F. (2020). The COVID-19 crisis: Opportunities for sustainable and proximity tourism. *Tourism Geographies, 22*(3), 690–694.

Salmela, T., Nevala, H., Nousiainen, M., & Rantala, O. (2021). Proximity tourism: A thematic literature review. *Matkailututkimus, 17*(1), 46–63.

Sigala, M. (2021). A bibliometric review of research on COVID-19 and tourism: Reflections for moving forward. *Tourism Management Perspectives, 40*, 100912.

SIGERIC. (2015). About us. https://www.sigeric.it/about-us/?lang=en

SIGERIC. (2020). Turismo in Lunigiana. Una Riflessione di fine estate. https://www.sigeric.it/turismo-in-lunigiana-una-riflessione-di-fine-estate/

Stylos, N. (2020). Technological evolution and tourist decision-making: A perspective article. *Tourism Review, 75*(1), 273–278.

Stylos, N. (2022a). An integrated duality theory framework (IDTF): Marking pathways for consumer decision-making researchers in the hospitality and tourism industry. *International Journal of Contemporary Hospitality Management, 34*(7), 2597–2619.

Stylos, N. (2022b). Big Data. In D. Buhalis (Ed.), *Encyclopedia of tourism management and marketing* (pp. 293–295). Edward Elgar Publishing.

Stylos, N., Fotiadis, A. K., Shin, D. D., & Huan, T. C. T. (2021a). Beyond smart systems adoption: Enabling diffusion and assimilation of smartness in hospitality. *International Journal of Hospitality Management, 98*. https://doi.org/10.1016/j.ijhm.2021.103042

Stylos, N., Zwiegelaar, J., & Buhalis, D. (2021b). Big data empowered agility for dynamic, volatile, and time-sensitive service industries: The case of tourism sector. *International Journal of Contemporary Hospitality Management, 33*(3), 1015–1036.

Terre del Magra. (2016). Lunigiana. https://terredelmagra.com/lunigiana/#:~:text=Politicamente%20la%20Lunigiana%20appartiene%20alla,bacino%20idrografico%20del%20fiume%20Magra

Terre di Lunigiana. (2022). La Lunigiana. https://www.terredilunigiana.com/lunigiana.php

Thrillist Travel. (2018). America's best small cities to move to before they get too popular. *Thrillist*, September 28.

Urry, J. (2002). *The tourist gaze* (2nd ed.). Sage.

Vassiliadis, C. A., & Stylos, N. (2022). Data mining. In D. Buhalis (Ed.), *Encyclopedia of tourism management and marketing* (pp. 790–792). Edward Elgar Publishing.

Visit Lunigiana. (2022). DOP & IGP products. https://visitlunigiana.it/what-to-taste/dop-igp-products/?lang=en

Visit Lunigiana. (2023). Who we are. https://visitlunigiana.it/?lang=en

Wiki Commons. (2008). A map of Lunigiana, Tuscany. https://commons.wikimedia.org/

Wiki Commons. (2012). Chestnut manual processing. https://commons.wikimedia.org/

Wiki Commons. (2014). Lunigiana semi-dilapidated castles. https://commons.wikimedia.org/

Zvaigzne, A., Litavniece, L., & Dembovska, I. (2022). Tourism seasonality: The causes and effects. *Worldwide Hospitality and Tourism Themes*.

Index

N. Stylos et al. (eds.), *Contemporary Marketing Management for Tourism and Hospitality*, https://doi.org/10.1007/978-3-031-65049-9

GPSR Compliance

The European Union's (EU) General Product Safety Regulation (GPSR) is a set of rules that requires consumer products to be safe and our obligations to ensure this.

If you have any concerns about our products, you can contact us on ProductSafety@springernature.com

In case Publisher is established outside the EU, the EU authorized representative is:

Springer Nature Customer Service Center GmbH
Europaplatz 3
69115 Heidelberg, Germany

The manufacturer's authorised representative in the EU is Springer Nature Customer Service Centre GmbH, Europaplatz 3, 69115 Heidelberg, Germany. If you have any concerns regarding our products, please contact ProductSafety@springernature.com

Printed and bound by CPI Group (UK) Ltd, Croydon, CR0 4YY
30/09/2026
02225595-0009